Shock Wave and High Pressure Phenomena

The Springer book series on Shock Wave and High Pressure Phenomena comprises monographs and multi-author volumes containing either original material or reviews of subjects within the field. All states of matter are covered. Methods and results of theoretical and experimental research and numerical simulations are included, as are applications of these results. The books are intended for graduate-level students, research scientists, mathematicians, and engineers. Subjects of interest include properties of materials at both the continuum and microscopic levels, physics of high rate deformation and flow, chemically reacting flows and detonations, wave propagation and impact phenomena.

The following list of subject areas further delineates the purview of the series. In all cases entries in the list are to be interpreted as applying to nonlinear wave propagation and high pressure phenomena. Development of experimental methods is not identified specifically, being regarded as a normal part of research in all areas of interest.

Material Properties

- Equation of state including chemical and phase composition, ionization, etc.
- Constitutive equations for inelastic deformation
- Fracture and fragmentation
- Dielectric and magnetic properties
- Optical properties and radiation transport
- Metallurgical effects
- Spectroscopy

Physics of Deformation and Flow

- Dislocation physics, twinning, and other microscopic deformation mechanisms
- Shear banding
- Mesoscale effects in solids
- Turbulence in fluids
- Microfracture and cavitation

Explosives

- Detonation of condensed explosives and gases
- Initiation and growth of reaction
- Detonation wave structures
- Explosive materials

Wave Propagation and Impact Phenomena in Solids

- Shock and decompression wave propagation
- Shock wave structure
- Penetration mechanics

Gasdynamics

- Chemically Reacting Flows
- Blast waves
- Multiphase flow

Numerical Simulation and Mathematical Theory

- Mathematical methods
- Wave propagation codes
- Molecular dynamics

Applications

- Material modification and synthesis
- Military ordnance
- Geophysics and planetary science
- Medicine
- Aerospace and industrial applications
- Protective materials and structures
- Mining

Henry Foust

Speed of Sound and Critical Flow

Springer

Henry Foust
Department of Computer Science
and Engineering Technology
University of Houston-Downtown
Houston, TX, USA

ISSN 2197-9529 ISSN 2197-9537 (electronic)
Shock Wave and High Pressure Phenomena
ISBN 978-981-97-5917-0 ISBN 978-981-97-5918-7 (eBook)
https://doi.org/10.1007/978-981-97-5918-7

This Springer imprint is published by the registered company Springer Nature Singapore Pte Ltd.
The registered company address is: 152 Beach Road, #21-01/04 Gateway East, Singapore 189721, Singapore

Preface

This book provides a primer on speed of sound through liquids, gases, and solids; it goes on to discuss a related phenomenon, which is critical flow. Critical flow is the idea that a pressurized vessel/conduit has an opening and we'd like to know the mass efflux from this opening? This question becomes challenging when the substance is no longer a pure gas but a mixture of liquid/gas or a mixture of substances.

In Chap. 1, an introduction to waves is considered. This introduction to waves includes a primer on vibration and speed of sound through gases, liquids, and solids.

In Chap. 2, an introduction to gas dynamics is provided. This introduction provides a basis for isothermal, isentropic, and choked (critical) gas flow. The chapter goes on to provide general conservation principles for a pure gas and ends with conservation principles for two-phase gas flows.

The speed of sound in pure gases is considered in Chap. 3. This starts with ideal, perfect gases and goes on to cubic equation of state gases to include van der Waals and Redlich–Kwong gases; the chapter ends with showing when a gas in terms of speed of sound can be considered ideal versus cubic and when varying specific heat is important.

Gas/liquid speed of sound is considered in Chap. 4. This section starts with the work of Karplus who considered water with air bubbles and goes on to consider the effects of surface tension on a gas/liquid system, mixtures of gases, and the speed of sound associated with a gas/liquid when we assume homogenous flow using the Omega method attributed to Leung.

The speed of sound in liquids is considered in Chap. 5. This is done by presenting Tait's equation of state for liquids and then determining the isothermal speed of sound based on Tait's equation and other variations considered. This chapter goes on to look at the speed of sound for liquid mixtures in a fashion similar to corresponding states, which is the framework for the Lee-Kesler compressibility charts.

Speed of sound through solids is considered in Chap. 6 with a complete derivation for the equations associated with primary waves, secondary waves, and speed of sound in solids for a one-dimensional solid; the chapter ends with a comparison of theory and experimental data for various metals.

Critical flow for a pure gas is considered in Chap. 7 and begins with an ideal, perfect gas. Van der Waals critical flow is then considered, and the chapter ends by showing some of the complexity of considering a liquid/gas critical flow by introducing the homogenous equilibrium model.

Chapter 8 considers critical flow for a mixture of liquid/gas; formally introduces homogenous equilibrium models; frozen equilibrium models; separated flow models; and goes on to provide a discussion of the non-equilibrium models of Henry and Fauske. This chapter ends with a brief review of the work on non-equilibrium critical flow since 1971.

In Chap. 9, a gas/liquid homogenous critical flow is considered and based on the Omega method attributed to Leung. Flows associated with nozzles and pipes are considered for sub-cooled fluids, non-condensable gases and go on to consider critical flow for mixtures of substances and non-ideal conditions.

Houston, USA Henry Foust

Contents

Chapter 1
Fundamentals of Waves

Abstract In this chapter, a brief discussion is made for both longitudinal and transverse waves and its connection with vibration; a primer of vibration is also given. Fundamental equations for the speed of sound through liquids, gases, and solids are provided.

Keywords Waves · Vibrations · Traverse waves · Longitudinal waves · Young's modulus

1.1 Preview

Vibration is a periodic motion with one particular model given in Fig. 1.1.

More formally vibration is defined as "… a mechanical phenomenon whereby oscillations occur about an equilibrium point [1]" and vibrations form a class of mechanics known as wave theory, where "… a **wave** is a propagating dynamic disturbance (change from equilibrium) of one or more quantities. Waves can be periodic, in which case those quantities oscillate repeatedly about an equilibrium (resting) value at some frequency [2]."

"**Acoustics** is a branch of physics that deals with the study of mechanical waves in gases, liquids, and solids including topics such as vibration, sound, ultrasound, and infrasound [3]."

A mathematical description of vibration is given below.

The conservation of momentum for Fig. 1.1 is

$$\rightarrow_{+} \sum f_x = -kx - c\dot{x} + f(t) = m\frac{d^2x}{dt^2}, \tag{1.1}$$

where k is the stiffness of the spring, c is a constant associated with dampening, and m is the mass of the object.

Or

H. Foust, *Speed of Sound and Critical Flow*, Shock Wave and High Pressure Phenomena, https://doi.org/10.1007/978-981-97-5918-7_1

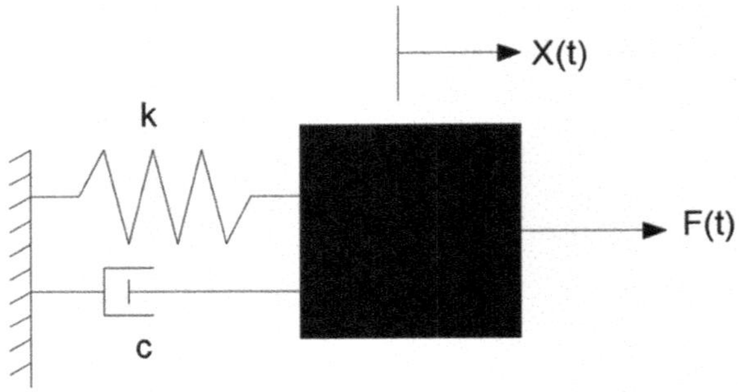

Fig. 1.1 Forced, damped mass-spring system

$$\frac{d^2x}{dt^2} + \frac{k}{m}x + c\dot{x} = \frac{d^2x}{dt^2} + \omega_0^2 x + \frac{c}{m}\dot{x} = \frac{f(t)}{m}. \tag{1.2}$$

And the natural frequency of this system is

$$\omega_n = \sqrt{\frac{k}{m}}. \tag{1.3}$$

A solution to Eq. 1.2 is

$$x = A_1 \cos(\varpi_0 t) + A_2 \sin(\varpi_0 t), \tag{1.4}$$

where A_1 and A_2 are determined by the initial conditions and can be shown to be complex conjugates [4–6].

It can be shown that vibration systems are a particular type of wave [4–6] and that a whole class of systems will have a more general solution given as

$$y = y_1(ct - x) + y_2(ct + x), \tag{1.5}$$

where $\{y_1, y_2\}$ are particular solutions, c is the wave speed, x is position, and t is time.

One practical application of wave theory and acoustics is sound. **Sound** [4, 6] is a particular periodic motion that occurs within an audible frequency range with a possible analogy being molecules as spheres that are connect by springs and the "natural frequency" or speed of the system is proportion to the square root of the stiffness (k) over the mass (m) or

$$c \propto \sqrt{\frac{k}{m}}. \tag{1.6}$$

For **solids** (thin rods—essentially one dimensional), we'll see this speed is

$$c \approx \sqrt{\frac{E}{\rho}}, \tag{1.7}$$

where E is Young's modulus, which is a measure of stiffness (resistance to elastic change under normal stresses), and ρ is the density of the solid.

And for **liquids**

$$c \approx \sqrt{\frac{K_s}{\rho}}, \tag{1.8}$$

where K_s is the bulk modulus along an isentropic path and ρ is the density of the liquid.

For **gases** when they're ideal

$$PV = nR_u T, \quad P = R_u \rho T \tag{1.9}$$

or specific to the type of gas

$$P = R_g \rho T. \tag{1.10}$$

And by definition, which will be proven below,

$$c^2 = \left(\frac{\partial P}{\partial \rho}\right)_s = \gamma \left(\frac{\partial P}{\partial \rho}\right)_T = \gamma R_g T, \tag{1.11}$$

where the speed of sound is typically defined along an isentropic path, γ is the thermodynamic constant, R_g is the gas constant particular to the gas, and T is the temperature.

The challenge is that for solids the geometry and material properties of the object under consideration affects the wave speed [4, 6] and for liquids and gases the states (temperature, pressure, density) can have an effect on the speed of sound; additionally, we'll see that the composition of the liquid or gas can have a profound effect on the speed of sound, i.e., a liquid–gas flow will have a speed of sound less than the speed of sound for the pure liquid or the pure gas form.

This chapter will become the basis to the discussion of speed of sound through gases (Chaps. 3 and 4), through liquids (Chap. 5), and through solids (Chap. 6). Chapter 2 will become the basis to the discussion of critical flow for one phase liquids (Chap. 7), two-phase liquids non-homogenous (Chap. 8), and two-phase liquids, homogenous (Chap. 9).

1.2 Introduction to Vibrations

In this section, a brief review will be given of vibrations that includes the development of a second order differential equation with a possible forcing function and dampening that is a mathematical model of an idealization, the mass-spring system.

1.2.1 Transverse and Longitudinal Waves

A **transverse wave** is one where the wave motion is perpendicular to the direction of travel. Examples of transverse waves include vibrating strings, water surface waves, electromagnetic waves, and seismic S waves (see Fig. 1.2).

A **longitudinal wave** is one where the wave motion is in the same direction as the direction of travel. Examples of longitudinal waves include waves in springs, sound waves, tsunami waves, and seismic P waves (see Fig. 1.2).

We'll see that the wave speed for seismic S and P waves can be described by

Secondary wave speed is

$$c_s = \sqrt{\frac{G}{\rho}}. \tag{1.12}$$

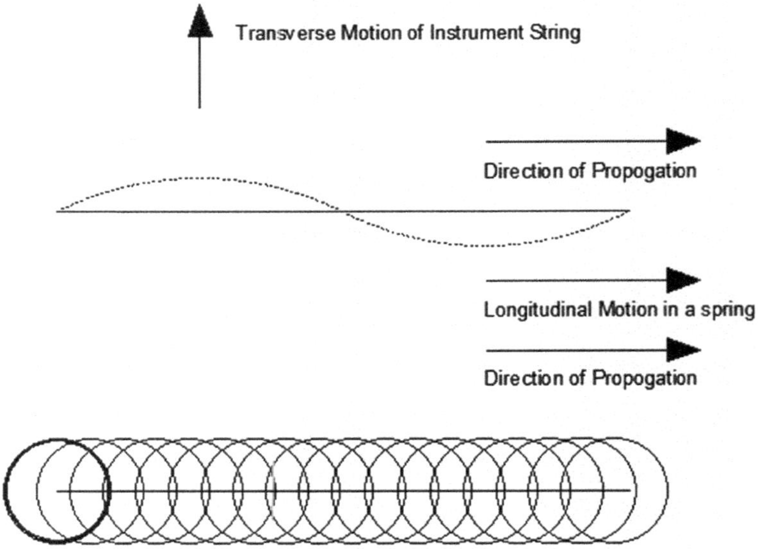

Fig. 1.2 Transverse and longitudinal waves

And primary wave speed is

$$c_p = \sqrt{\frac{E(1-\nu)}{\rho(1+\nu)(1-2\nu)}}, \tag{1.13}$$

where E is the Young's modulus, ρ is the density, ν is the Poisson ratio, and G is the modulus of rigidity and E, ν and G are related by

$$E = 2G(1+\nu). \tag{1.14}$$

Much of this section will utilize a spring-mass system to express idealized vibration systems that will be shown to be expressed by some form of

$$\ddot{x} + a\dot{x} + bx = f(x). \tag{1.15}$$

"Sound waves are longitudinal waves of compression and rarefaction in which the air molecules move back and forth parallel to the direction of wave travel centered on an average position, resulting in no net movement of the molecules [4, 6]." In Fig. 1.3, some of the differences between rarefaction and compression waves are shown along with connections to vibrations and waves.

1.2.2 Undamped, Mass-Spring System

Applying Newton's Second Law in the horizontal direction for Fig. 1.1 results in

$$\sum_{(+)\rightarrow} F_x = -kx = m\frac{d^2x}{dt^2}, \tag{1.16}$$

where k is the spring constant and m is the mass of the block.

Or

$$m\frac{d^2x}{dt^2} + kx = 0, \quad \ddot{x} - \omega_n^2 = 0, \tag{1.17}$$

where

$$\omega_n = \text{natural frequency of system} = \sqrt{\frac{k}{m}}. \tag{1.18}$$

We'll discuss solutions to Eq. 1.17 below.

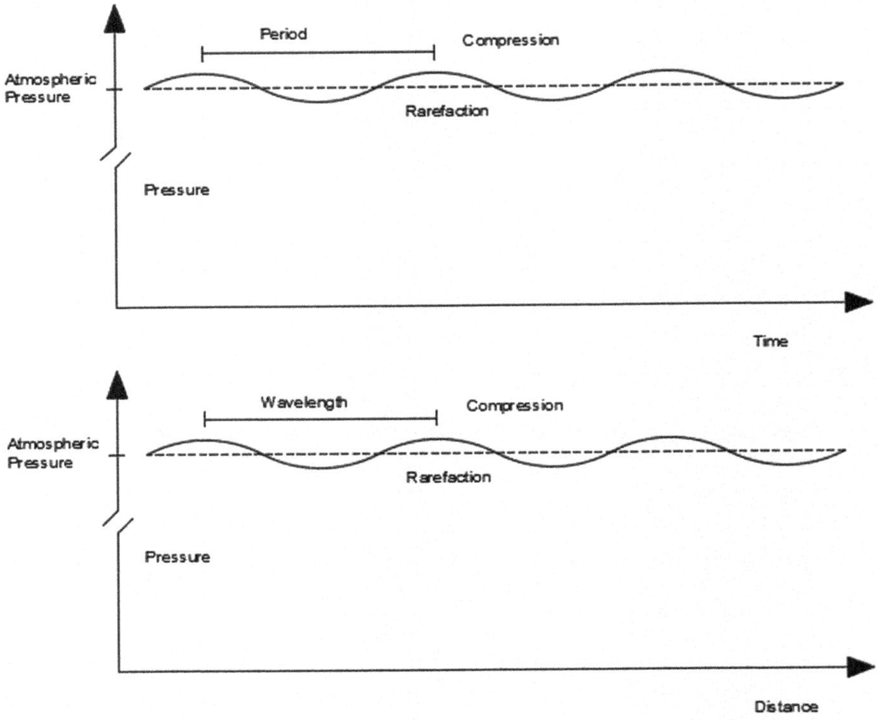

Fig. 1.3 Compression/rarefaction waves

1.2.3 Damped, Mass-Spring System

Applying Newton's Second Law in the horizontal direction for Fig. 1.4 results in

$$\sum_{(+)\rightarrow} F_x = -kx - c\dot{x} = m\frac{\mathrm{d}^2x}{\mathrm{d}t^2}, \tag{1.19}$$

where k is the spring constant, c is a constant associated with a damper, and m is the mass of the block.

Equation 1.19 can also be written as

$$m\ddot{x} + c\dot{x} + kx = 0 \tag{1.20}$$

or

$$\ddot{x} + \frac{c}{m} + \omega_{\mathrm{n}}^2 = 0 \tag{1.21}$$

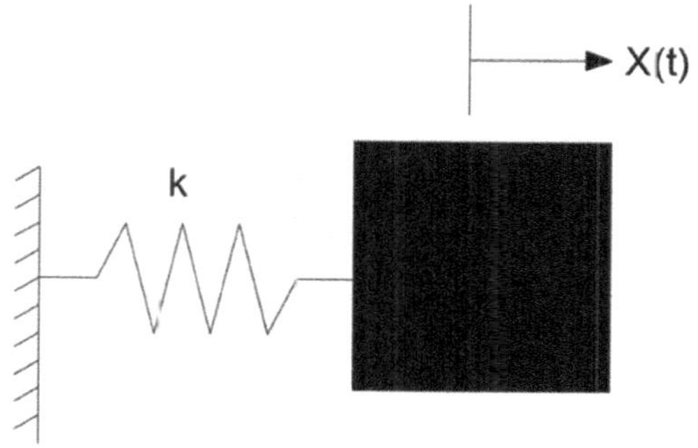

Fig. 1.4 Mass-spring system

1.2.4 Forced, Mass-Spring System

Applying Newton's Second Law in the horizontal direction for Fig. 1.5 results in

$$\sum_{(+)\rightarrow} F_x = -kx + f(t) = m\frac{d^2x}{dt^2}, \tag{1.22}$$

where k is the spring constant, c is a constant associated with a damper, and m is the mass of the block.

Equation 1.22 can also be written as

$$m\ddot{x} + kx = f(t) \tag{1.23}$$

or

$$\ddot{x} + \omega_n^2 = \frac{f(t)}{m} \tag{1.24}$$

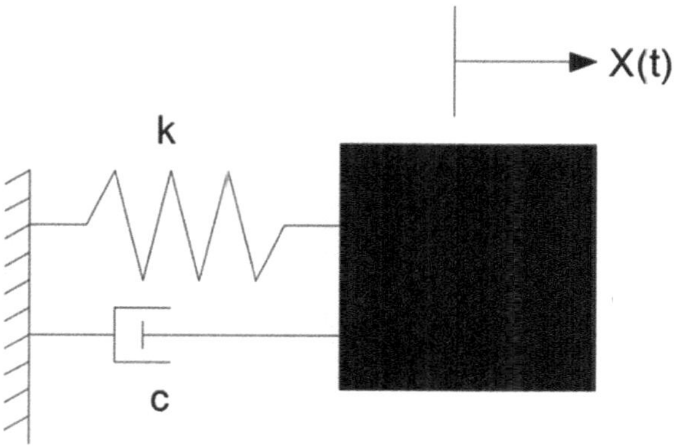

Fig. 1.5 Damped, mass-spring system

1.2.5 Forced, Damped Mass-Spring System

Applying Newton's Second Law in the horizontal direction for Fig. 1.6 results in

$$\sum_{(+)\rightarrow} F_x = -kx - c\dot{x} + f(t) = m\frac{\mathrm{d}^2 x}{\mathrm{d}t^2}, \tag{1.25}$$

where k is the spring constant, c is a constant associated with a damper, and m is the mass of the block.

Equation 1.25 can also be written as (Fig. 1.7)

$$m\ddot{x} + c\dot{x} + kx = f(t) \tag{1.26}$$

or

$$\ddot{x} + \frac{c}{m} + \omega_{\mathrm{n}}^2 = \frac{f(t)}{m}. \tag{1.27}$$

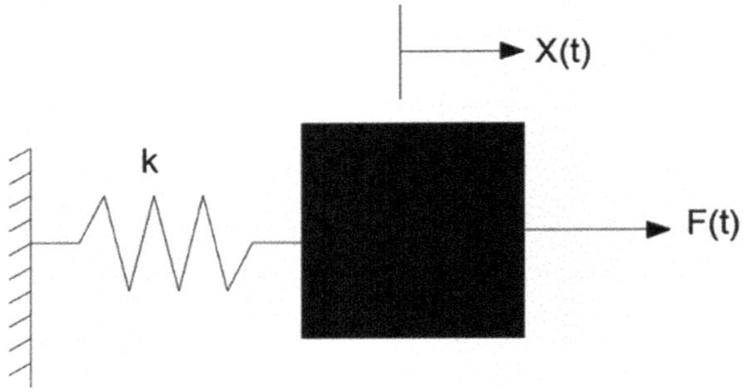

Fig. 1.6 Forced, dampened mass-spring system

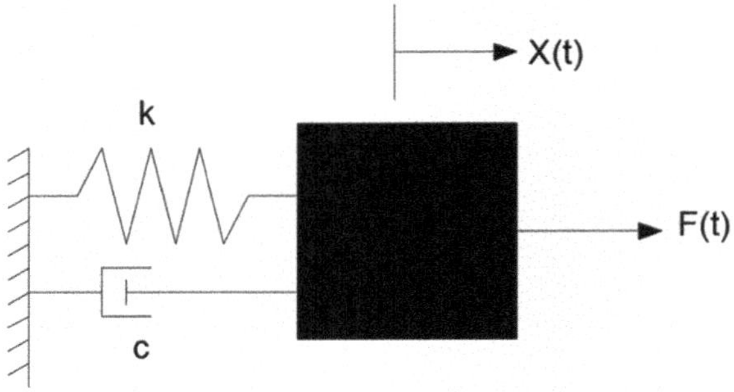

Fig. 1.7 Forced, mass-spring system

1.2.6 Mathematical Solutions of Various Mass-Spring Systems

Undamped, unforced mass-spring system …

Given

$$\ddot{x} + \omega_n^2 = 0. \tag{1.28}$$

It can be shown that the solution to Eq. 1.28 [4–6] is

$$x = A\sin(\omega_n t) + B\cos(\omega_n t), \tag{1.29}$$

where $\{A, B\}$ can be determined by initial conditions given as

$$x(t)' = A\omega_n\cos(\omega_n t) - B\omega_n\sin(\omega_n t). \tag{1.30}$$

And

$$x(t)'' = -A\omega_n^2\sin(\omega_n t) - B\omega_n^2\cos(\omega_n t). \tag{1.31}$$

The constants of integration $\{A, B\}$ can also be expressed in terms of sine and cosine such that

$$A = C\cos(\phi), \quad B = C\sin(\phi). \tag{1.32}$$

And substituting into Eq. 1.29 results in

$$x(t) = C\cos(\phi)\sin(\omega_n t) + C\sin(\phi)\cos(\omega_n t). \tag{1.33}$$

And from the trigonometric identity

$$\sin(a + b) = \sin(a)\cos(b) - \cos(a)\sin(b). \tag{1.34}$$

Results in

$$x(t) = C\sin(\omega_n t + \phi), \tag{1.35}$$

where ϕ is a phase shift and Eq. 1.35 is a solution to all over mass-spring systems considered.

Undamped, force mass-spring systems …

Given

$$m\ddot{x} + kx = F_0\sin(\omega_0 t) = f(t). \tag{1.36}$$

The solution has two parts

$$x(t) = x_c(t) + x_p(t), \tag{1.37}$$

where the complimentary solutions (x_c) are the case where the right-hand side of Eq. 1.36 is zero and is

$$x(t) = C\sin(\omega_n t + \phi). \tag{1.38}$$

And the particular solution (x_p) is

$$x_p = X\sin(\omega_0 t). \tag{1.39}$$

Taking the second derivative of Eq. 1.39 results in

$$x_p'' = -X\omega_0^2\sin(\omega_0 t). \tag{1.40}$$

And substituting Eqs. 1.39 and 1.40 into Eq. 1.36 results in

$$-X\omega_0^2\sin(\omega_0 t) + \omega_n^2\{X\sin(\omega_0 t)\} = \frac{F_0}{m}\sin(\omega_0 t). \tag{1.41}$$

Solving for X results in

$$X = \frac{F_0/m}{1 - \left(\frac{\omega_0}{\omega_n}\right)^2}. \tag{1.42}$$

Substituting this relationship into Eq. 1.39 provides the particular solution

$$x_p = \frac{F_0/m}{1 - \left(\frac{\omega_0}{\omega_n}\right)^2}\sin(\omega_0 t). \tag{1.43}$$

And the complete solution is

$$x = x_c + x_p = C\sin(\omega_n t + \phi) + \frac{F_0/m}{1 - \left(\frac{\omega_0}{\omega_n}\right)^2}\sin(\omega_0 t) \tag{1.44}$$

or

$$x(t) = x_c(t) + x_p(t) = \text{transient} + \text{steady state}. \tag{1.45}$$

And

$$t = \infty \rightarrow x(t) = x_p(t). \tag{1.46}$$

Damped mass-spring system ...
Given

$$m\ddot{x} + c\dot{x} + kx = 0, \tag{1.47}$$

where the solution to be utilized is

$$x(t) = e^{\lambda t}. \tag{1.48}$$

The first and second derivatives of Eq. 1.48 are

$$x(t)' = \lambda e^{\lambda t}, \quad x(t)'' = \lambda^2 e^{\lambda t}. \tag{1.49}$$

Substitution Eq. 1.49 into Eq. 1.47 results in

$$e^{\lambda t}\{m\lambda^2 + c\lambda + k\} = 0 \tag{1.50}$$

or

$$m\lambda^2 + c\lambda + k = 0, \tag{1.51}$$

where solutions to λ are

$$\lambda_1 = \frac{c}{2m} + \sqrt{\left(\frac{c}{2m}\right)^2 - \frac{k}{m}}. \tag{1.52}$$

And

$$\lambda_2 = \frac{c}{2m} - \sqrt{\left(\frac{c}{2m}\right)^2 - \frac{k}{m}}. \tag{1.53}$$

The determinant of Eqs. 1.52 and 1.53 provides some qualitative results of the system.

When the determinant shown in Eqs. 1.52 and 1.53 is equal to zero, then the system is considered overdamped such that

$$\left(\frac{c_{\text{critical}}}{2m}\right)^2 - \frac{k}{m} = 0. \tag{1.54}$$

Solving for c_{critical} results in

$$c_{\text{critical}} = 2m\sqrt{\frac{k}{m}} = 2m\omega_n, \tag{1.55}$$

where ω_n is the natural frequency of the system.

When $c = c_{critical}$, then the system is *critically damped*, and this is a situation where the vibration can be detrimental to the system.

When $c > c_{critical}$, then the system is considered *over-damped*. This is a situation where both roots are real and there is essentially no vibration.

And when $c < c_{critical}$, then the system is considered *under-damped*. The roots are complex and vibration results but dampen with time.

1.2.7 Resonance

Damped, forced mass-spring system

Given

$$m\ddot{x} + c\dot{x} + kx = F_0 \sin(\omega_0 t). \tag{1.56}$$

Let's consider a particular solution, which is the steady state part of the solution to Eq. 1.56 and is given as

$$x_p = X' \sin(\omega_0 t - \phi'). \tag{1.57}$$

The first and second derivations of this solution are

$$x_p' = X'^{\omega_0} \cos(\omega_0 t - \phi'), \quad x_p'' = -X'^{\omega_0^2} \sin(\omega_0 t - \phi'). \tag{1.58}$$

Substitution of Eq. 1.58 into Eq. 1.56 results in

$$-X' m\omega_0^2 \sin(\omega_0 t - \phi') + X' c\omega_0 \cos(\omega_0 t - \phi') + X' k \sin(\omega_0 t - \phi') = F_0 \sin(\omega_0 t). \tag{1.59}$$

When $(\omega_0 t - \phi' = 0)$

$$X' c\omega_0 = F_0 \sin(\phi'). \tag{1.60}$$

When $(\omega_0 t - \phi' = \pi/2)$

$$-X' m\omega_0^2 + X' k = F_0 \cos(\phi'). \tag{1.61}$$

If we square Eqs. 1.60 and 1.61 and add together the result is

$$(X' c\omega_0)^2 + (-X' m\omega_0^2 + X' k)^2 = F_0^2. \tag{1.62}$$

Solving Eq. 1.62 for X' results in

$$X' = \frac{F_0}{\sqrt{\left(k - m\omega_0^2\right)^2 + (c\omega_0)^2}}. \quad (1.63)$$

And dividing Eq. 1.60 by Eq. 1.61 results in

$$\tan(\phi') = \frac{c\omega_0}{k - m\omega_0^2}. \quad (1.64)$$

A magnitude factor (MF) has been defined as

$$\mathrm{MF} = \frac{X'}{F_0/k} = \frac{1}{\sqrt{\left[1 - (Q)^2\right]^2 + \left[2c'Q\right]^2}}, \quad (1.65)$$

where Q is

$$Q = \frac{\omega_0}{\omega_\mathrm{n}}. \quad (1.66)$$

And c' is

$$c' = \frac{c}{c_\mathrm{c}}. \quad (1.67)$$

A graph is given for MF versus Q, and it is seen that at Q equal to 1 MF increases without bounds when c' approaches zero, and this is the basis to the idea of resonance. When the frequency of the forcing function equals the natural frequency (ω_n), problems can develop (Fig. 1.8).

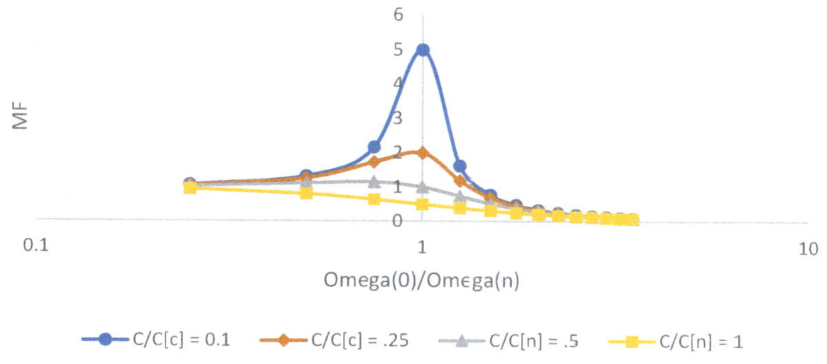

Fig. 1.8 MP versus Q

1.3 Waves Speed in Solids

Given the following system where T is tension and the string has a unit weight of ρ' (see Fig. 1.9).

While making the following assumptions

1. The tension (T) is constant and independent of position.
2. The angle with the horizon, θ_1 and θ_2, is very small.
3. The motion is only transverse—in the same direction as T.
4. Ignored friction, stiffness, and gravity.

Thus, there are unbalanced forces in both the x-direction

$$-T\cos(\theta_1) + T\cos(\theta_2) = f_x. \tag{1.68}$$

And in the y-direction

$$-T\sin(\theta_1) + T\sin(\theta_2) = f_y. \tag{1.69}$$

And due to the very small angle

$$f_x \approx 0. \tag{1.70}$$

And

$$\tan(\theta_i) = \frac{\partial y_i}{\partial x}. \tag{1.71}$$

Substitution of Eq. 1.71 into Eq. 1.69 results in

$$-T\frac{\partial y_1}{\partial x} + T\frac{\partial y_2}{\partial x} = f_y. \tag{1.72}$$

The unbalance force, f_y, is itself equal to the change in momentum

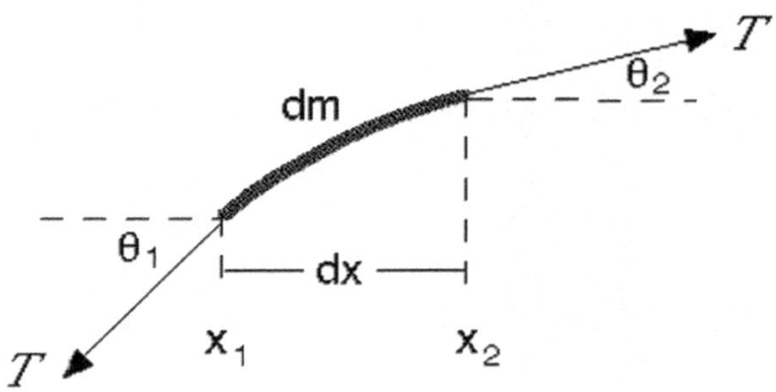

Fig. 1.9 String segment [4, 6]

$$m\frac{\partial^2 y}{\partial t^2} = \rho' \Delta x \frac{\partial^2 y}{\partial t^2}. \tag{1.73}$$

Substitution of Eq. 1.73 into Eq. 1.72 itself results in

$$-T\frac{\partial y_1}{\partial x} + T\frac{\partial y_2}{\partial x} = f_y = \rho \Delta x \frac{\partial^2 y}{\partial t^2}. \tag{1.74}$$

Diving both sides by Δx and taking the limit as $\Delta x \to 0$ results in

$$T\frac{\partial^2 y}{\partial x^2} = \rho' \frac{\partial^2 y}{\partial t^2} \tag{1.75}$$

or

$$\frac{\partial^2 y}{\partial t^2} = c^2 \frac{\partial^2 y}{\partial x^2}, \tag{1.76}$$

where

$$c = \text{Transverse wave speed} = \sqrt{\frac{T}{\rho'}}. \tag{1.77}$$

It can be shown [4, 6] that for a thin rod, which is essentially one dimensional, the longitudinal wave speed is

$$c = \sqrt{\frac{E}{\rho}}, \tag{1.78}$$

where E is the Young's modulus for the material.

1.4 Wave Speed in Liquids

We've seen for solids the speed of sound can take the form

$$c^2 \approx \sqrt{\frac{E}{\rho}}, \tag{1.79}$$

where E is the Young's modulus of the material and ρ is the material's density.

If we define the strain [4, 6] as

$$\theta = \frac{\Delta V}{V}. \tag{1.80}$$

And theta can be shown to be directly related to the strain in the three principle directions

$$\begin{aligned}\theta &= \frac{\Delta V}{V} = \frac{\left(\Delta x + \frac{\partial \varepsilon}{\partial x}\Delta x\right)\left(\Delta y + \frac{\partial \eta}{\partial y}\Delta y\right)\left(\Delta z + \frac{\partial \zeta}{\partial z}\Delta z\right) - \Delta x \Delta y \Delta z}{\Delta x \Delta y \Delta z} \\ &= \frac{\partial \varepsilon}{\partial x} + \frac{\partial \eta}{\partial y} + \frac{\partial \zeta}{\partial z} = \epsilon_{xx} + \epsilon_{yy} + \epsilon_{zz}\end{aligned} \tag{1.81}$$

where the elongation on a rod for an isotropic material is

$$\theta_{\text{rod}} = \epsilon_{xx}(1 - 2\sigma). \tag{1.82}$$

Since $\epsilon_{yy} = \epsilon_{zz} = -\sigma\epsilon_{xx}$ and the strains are related to the stresses $\{f_{xx}, f_{yy}, f_{zz}\}$ through

$$\begin{bmatrix} \epsilon_{xx} \\ \epsilon_{yy} \\ \epsilon_{zz} \end{bmatrix} = \begin{bmatrix} \frac{1}{E} & -\frac{\sigma}{E} & -\frac{\sigma}{E} \\ -\frac{\sigma}{E} & \frac{1}{E} & -\frac{\sigma}{E} \\ -\frac{\sigma}{E} & -\frac{\sigma}{E} & \frac{1}{E} \end{bmatrix} \begin{bmatrix} f_{xx} \\ f_{yy} \\ f_{zz} \end{bmatrix}. \tag{1.83}$$

The terms on the left-hand side add to θ and the terms on the right-hand side add to

$$\frac{1}{E}\left[f_{xx} + f_{yy} + f_{zz}\right] - 2\frac{\sigma}{E}\left[f_{xx} + f_{yy} + f_{zz}\right], \tag{1.84}$$

where the definition of pressure (P) is given as

$$P = -\frac{1}{3}\left[f_{xx} + f_{yy} + f_{zz}\right]. \tag{1.85}$$

Substitution of Eq. 1.85 into Eq. 1.84 results in

$$\theta = -3\frac{P}{E} - 6\frac{P}{E}\sigma = -3\frac{P}{E}(1 - 2\sigma). \tag{1.86}$$

For a fluid,

$$P = -K_s\theta, \tag{1.87}$$

where K_s is the bulk modulus (isentropic) for that fluid and substituting Eq. 1.87 into Eq. 1.86 results in

$$\theta = 3\frac{K_s\theta}{E}(1 - 2\sigma), \tag{1.88}$$

or

$$E = 3K_s(1 - 2\sigma), \tag{1.89}$$

establishing a relationship between Young's modulus (E) and bulk modulus (K_s).

Again the dilation [4] is

$$\theta = \frac{\Delta V}{V} = \frac{\partial \varepsilon}{\partial x} + \frac{\partial \eta}{\partial y} + \frac{\partial \zeta}{\partial z} = \nabla \cdot \varrho. \tag{1.90}$$

The dilation can be written compactly as

$$\theta = \nabla \cdot \varrho. \tag{1.91}$$

And so pressure can be defined as

$$P = -K_s\theta = -K_s\nabla \cdot \varrho. \tag{1.92}$$

Given a cube of fluid that experiences stress due to wave pressure variations, along the x-axis the force on face $ABCD$ is

$$F_{ABCD} = P\Delta y\Delta z. \tag{1.93}$$

And along the x-axis on face $EFGH$ the force is

$$F_{EFGH} = \left\{P + \left(\frac{\partial P}{\partial x}\Delta x\right)\right\}\Delta y\Delta z. \tag{1.94}$$

Utilizing Newton's Second Law along the x-axis, the result is

$$\Delta F_x = -\frac{\partial P}{\partial x}\Delta y\Delta z. \tag{1.95}$$

Along the y-axis, the result is

$$\Delta F_y = -\frac{\partial P}{\partial y}\Delta x\Delta z. \tag{1.96}$$

And along the z-axis, the result is

$$\Delta F_z = -\frac{\partial P}{\partial z}\Delta x\Delta y. \tag{1.97}$$

The net force in the cube is

$$\Delta F = -\nabla P\Delta x\Delta y\Delta z, \tag{1.98}$$

where

$$\nabla P = \hat{i}\frac{\partial P}{\partial x} + \hat{j}\frac{\partial P}{\partial y} + \hat{k}\frac{\partial P}{\partial z}. \tag{1.99}$$

The momentum change associated with Newton's Second Law is

$$\rho_0 \Delta x \Delta y \Delta z \frac{\partial^2 \varrho}{\partial t^2} = \Delta F = -\nabla P \Delta x \Delta y \Delta z, \tag{1.100}$$

or

$$\rho_0 \frac{\partial^2 \varrho}{\partial t^2} = -\nabla P. \tag{1.101}$$

Taking the divergence of both sides of Eq. 1.101 results in

$$-\nabla \cdot \nabla P = \rho_0 \frac{\partial^2 (\nabla \cdot \varrho)}{\partial t^2}. \tag{1.102}$$

Substitution of Eq. 1.102 into 1.100 results in

$$-\nabla \cdot \nabla P = \rho_0 \frac{\partial^2 (-P/K_s)}{\partial t^2}, \tag{1.103}$$

or

$$\nabla^2 P = \frac{1}{c_f^2}\frac{\partial^2 P}{\partial t^2}, \tag{1.104}$$

where

$$c_f^2 = \frac{K_s}{\rho_0}. \tag{1.105}$$

1.5 Wave Speed in Gases

Expanding pressure (P) as a function of density (ρ) for a perfect gas results in

$$P = P_0 + \left(\frac{\partial P}{\partial \rho}\right)(\rho - \rho_0) + 1/2\left(\frac{\partial P}{\partial \rho}\right)^2 (\rho - \rho_0)^2 + \cdots. \tag{1.106}$$

And defining s, A, and B as

$$s = \frac{\rho}{\rho_0} - 1, \quad A = \rho_0\left(\frac{\partial P}{\partial \rho}\right), \quad B = \rho_0^2\left(\frac{\partial P}{\partial \rho}\right)^2. \tag{1.107}$$

Substituting Eq. 1.107 into Eq. 1.106 results in

$$P - P_0 = As + \frac{1}{2Bs^2} + \cdots . \tag{1.108}$$

Retaining just the linear terms results in

$$P - P_0 = As. \tag{1.109}$$

By definition, A is the adiabatic bulk modulus (K_s) such that

$$A = K_s = \rho_0 \left(\frac{\partial P}{\partial \rho} \right)_s . \tag{1.110}$$

And for liquids and gases

$$c^2 = \left(\frac{\partial P}{\partial \rho} \right)_s = \frac{K_s}{\rho_0}. \tag{1.111}$$

Assuming an adiabatic relationship where

$$P v^\gamma = k \tag{1.112}$$

or

$$P = k \rho^\gamma . \tag{1.113}$$

And using the definition of speed of sound previously derived

$$\left(\frac{\partial P}{\partial \rho} \right)_s = c^2, \tag{1.114}$$

where the derivative of Eq. 1.114 takes the form

$$\frac{dP}{d\rho} = \gamma k \rho^{\gamma - 1} = \gamma \frac{P}{\rho} = c^2. \tag{1.115}$$

Equation 1.115 is appropriate for ideal, perfect gases.

References

1. https://en.wikipedia.org/wiki/Vibration. Accessed 7 May 2024
2. https://en.wikipedia.org/wiki/Wave. Accessed 7 May 2024
3. https://en.wikipedia.org/wiki/Acoustics. Accessed 19 Sept 2023

4. W.C. Elmore, M.A. Heald, *Physics of Waves* (Dover Books, 1985)
5. R.C. Hibbeler, *Dynamics*, 14th edn. (Pearson, Hoboken, NJ, 2016)
6. L.E. Kinsler, A.R. Frey, A.B. Coppens, J.V. Sanders, *Fundamentals of Acoustics*, 13th edn. (Wiley, New York, NY, 1982)

Chapter 2
Fundamentals of Compressible Flow

Abstract In this chapter, equations are developed to relate the states—pressure, temperature, and density—in isothermal, adiabatic, and choked flows. The chapter goes on to provide the effects of area changes on gaseous states through a converging–diverging nozzle. A non-dimensional form of the conservation equations is developed for a constant cross-sectional and adiabatic gas. The chapter then ends by showing conservation principles for a two-phase flow and provides several simplifications.

Keywords Gaseous conservation principles · Isothermal flow · Adiabatic flow · Choked flow · Conservation principles for two-phase flows

2.1 Preview

The purpose of this chapter is to provide a necessary foundation in gas dynamics in terms of conservation principles to include conservation principles for a two-phase fluid (liquid and gas). This foundation will be utilized in all subsequent chapters.

2.2 Conservation Principles for Single Phase Flow

Much of this chapter will deal with compressible flow through straight, horizontal pipes of constant cross-sectional area.

2.2.1 Conservation of Mass

$$G = \frac{\dot{m}}{A} = \rho V = \text{constant}. \tag{2.1}$$

H. Foust, *Speed of Sound and Critical Flow*, Shock Wave and High Pressure Phenomena, https://doi.org/10.1007/978-981-97-5918-7_2

2.2.2 Conservation of Energy

$$\frac{\mathrm{d}P}{\rho} + g\mathrm{d}z + \mathrm{d}\left(\frac{V^2}{2}\right) + \delta q + \delta w = 0. \tag{2.2}$$

Assuming no change in elevation and no work results in

$$\frac{\mathrm{d}P}{\rho} + \frac{G^2}{\rho}\mathrm{d}\left(\frac{1}{\rho}\right) + \delta q = 0. \tag{2.3}$$

And

$$\frac{\mathrm{d}P}{\rho} + \frac{G^2}{\rho}\mathrm{d}\left(\frac{1}{\rho}\right) + 2\frac{f}{D}\left(\frac{G}{\rho}\right)^2 \mathrm{d}L = 0, \tag{2.4}$$

where P is pressure, G is mass flux, ρ is density, f is friction factor, D is the inside diameter of the pipe, and dL is a unit length.

2.3 Isothermal Flow

Assuming an ideal gas allows for

$$\rho = \frac{P}{R_{\mathrm{g}}T}. \tag{2.5}$$

And substituting Eq. 2.5 into Eq. 2.4 results in

$$R_{\mathrm{g}}T\frac{\mathrm{d}P}{P} + R_{\mathrm{g}}T\frac{G^2}{P}\mathrm{d}\left(\frac{R_{\mathrm{g}}T}{P}\right) + 2\frac{f}{D}\mathrm{d}L + \left(\frac{GR_{\mathrm{g}}T}{P}\right)^2 = 0. \tag{2.6}$$

And

$$\frac{\mathrm{d}P}{P} + R_{\mathrm{g}}TG^2\left[\frac{1}{P}\mathrm{d}\left(\frac{1}{P}\right) + 2\frac{f}{DP^2}\mathrm{d}L\right], \tag{2.7}$$

or

$$P\mathrm{d}P + R_{\mathrm{g}}TG^2\left[P\mathrm{d}\left(\frac{1}{P}\right) + 2\frac{f}{D}\mathrm{d}L\right] = P\mathrm{d}P + R_{\mathrm{g}}TG^2\left[\frac{1}{1/P}\mathrm{d}\left(\frac{1}{P}\right) + 2\frac{f}{D}\mathrm{d}L\right], \tag{2.8}$$

which upon integration yields

$$\frac{P_1^2 - P_2^2}{2} = R_g T G^2 \left[\ln(1/P_2) - \ln(1/P_1) + 2\frac{fL}{D} \right]. \quad (2.9)$$

And solving for G provides

$$G^2 = \frac{\frac{P_1^2 - P_2^2}{2R_g T}}{\frac{2fL}{D} + \ln\left(\frac{P_1}{P_2}\right)}. \quad (2.10)$$

If the ratio of pressure, $\frac{P_1}{P_2}$, is to be determined for a known P_1, then another form of Eq. 2.9 is

$$\left(\frac{P_2}{P_1}\right)^2 = 1 - 2R_g T\left(\frac{G}{P_1}\right)^2 \left[2\frac{fL}{D} - \ln\left(\frac{P_2}{P_1}\right)\right], \quad (2.11)$$

where the friction factor can be determined from

$$\frac{1}{\sqrt{f}} = -4\log\left[\frac{\epsilon/D}{3.7}\right]. \quad (2.12)$$

2.4 Adiabatic Flow

Given the following adiabatic relationships.

$$\frac{\rho}{\rho_1} = \left(\frac{P}{P_1}\right)^{1/\gamma}, \quad \frac{T}{T_1} = \left(\frac{P}{P_1}\right)^{\gamma - 1/\gamma}. \quad (2.13)$$

And

$$\frac{dP}{\rho} + \frac{G^2}{\rho} d\left(\frac{1}{\rho}\right) + 2\frac{f}{D}\left(\frac{G}{\rho}\right)^2 dL = 0. \quad (2.14)$$

Substituting Eq. 2.13 into Eq. 2.14 results in

$$\frac{dP}{\rho_1\left(\frac{P}{P_1}\right)^{1/\gamma}} + \frac{G^2}{\rho_1\left(\frac{P}{P_1}\right)^{1/\gamma}} d\left(\frac{1}{\rho_1\left(\frac{P}{P_1}\right)^{1/\gamma}}\right) + 2\frac{f}{D}\left(\frac{G}{\rho_1\left(\frac{P}{P_1}\right)^{1/\gamma}}\right)^2 dL = 0 \quad (2.15)$$

or

$$\rho_1\left(\frac{P}{P_1}\right)^{1/\gamma}\mathrm{d}P + G^2\rho_1\left(\frac{P}{P_1}\right)^{1/\gamma}\mathrm{d}\left(\frac{1}{\rho_1\left(\frac{P}{P_1}\right)^{1/\gamma}}\right) + 2\frac{f}{D}G^2\mathrm{d}L = 0. \tag{2.16}$$

And

$$\frac{\rho_1}{P_1^{1/\gamma}}P^{1/\gamma}\mathrm{d}P + G^2\frac{1}{1/P^{1/\gamma}}\mathrm{d}\left(\frac{1}{P^{1/\gamma}}\right) + 2\frac{f}{D}G^2\mathrm{d}L = 0. \tag{2.17}$$

Upon integration from P_1 to P_2 results in

$$\frac{\gamma}{\gamma+1}\frac{\rho_1}{P_1^{1/\gamma}}\left[P_2^{\frac{\gamma+1}{\gamma}} - P_1^{\frac{\gamma+1}{\gamma}}\right] - \frac{G^2}{\gamma}\ln\left(\frac{P_2}{P_1}\right) = -2\frac{f}{D}G^2L. \tag{2.18}$$

And

$$G^2\left[\frac{1}{\gamma}\ln\left(\frac{P_2}{P_1}\right) - 2\frac{f}{D}L\right] = \frac{\gamma}{\gamma+1}\frac{\rho_1}{P_1^{1/\gamma}}\left[P_2^{\frac{\gamma+1}{\gamma}} - P_1^{\frac{\gamma+1}{\gamma}}\right]. \tag{2.19}$$

And solving for G^2 results in

$$G^2 = \frac{\frac{\gamma}{\gamma+1}\frac{\rho_1}{P_1^{1/\gamma}}\left[P_2^{\frac{\gamma+1}{\gamma}} - P_1^{\frac{\gamma+1}{\gamma}}\right]}{\left[\frac{1}{\gamma}\ln\left(\frac{P_2}{P_1}\right) - 2\frac{f}{D}L\right]} \tag{2.20}$$

or

$$G^2 = \frac{\frac{\gamma}{\gamma+1}\rho_1 P_1\left[\left(\frac{P_2}{P_1}\right)^{\frac{\gamma+1}{\gamma}} - 1\right]}{\left[\frac{1}{\gamma}\ln\left(\frac{P_2}{P_1}\right) - 2\frac{f}{D}L\right]}. \tag{2.21}$$

2.5 Choked Flow

From the conservation of mass,

$$G = \frac{\dot{m}}{A} = \rho V. \tag{2.22}$$

And the derivative $\partial G/\partial P$ set equal to zero maximizes the mass flow and is given as

$$\frac{\partial G}{\partial P} = \rho \frac{\partial V}{\partial P} + V \frac{\partial \rho}{\partial P} = 0. \tag{2.23}$$

From the conservation of energy and for adiabatic flow,

$$\frac{\mathrm{d}P}{\rho} + V\mathrm{d}V = 0, \quad \frac{\partial V}{\partial P} = -\frac{1}{\rho V}. \tag{2.24}$$

Substitution of Eq. 2.24 into Eq. 2.23 is

$$\frac{-1}{V} + V \frac{\partial \rho}{\partial P} = 0. \tag{2.25}$$

And by definition, the speed of sound along an isentropic path is

$$\left(\frac{\partial P}{\partial \rho}\right)_s = c^2. \tag{2.26}$$

Substitution of Eq. 2.26 into Eq. 2.25 results in

$$\frac{-1}{V} + \frac{V}{c^2} = 0, \quad V = c. \tag{2.27}$$

Thus, the mass velocity (G) is maximized when the velocity at the choke point is the local speed of sound. This result is only valid for a pure substance, and we'll see when we deal with two-phase components that the velocity at the choke point for a mixture of liquid and gas will be below the speed of sound for either the pure liquid or the pure gas.

2.6 Area Change Effects

From conservation of mass [1, 2]

$$\rho VA = \text{constant}. \tag{2.28}$$

And

$$\ln(\rho) + \ln(V) + \ln(A) = \text{constant}. \tag{2.29}$$

Taking the derivative of both sides

$$\frac{\mathrm{d}\rho}{\rho} + \frac{\mathrm{d}V}{V} + \frac{\mathrm{d}A}{A} = 0. \quad (2.30)$$

Substituting Eq. 2.25 for dV results in

$$\frac{\mathrm{d}\rho}{\rho} + \frac{-\mathrm{d}P}{\rho V^2} + \frac{\mathrm{d}A}{A} = 0. \quad (2.31)$$

And solving for dP/ρ gives us

$$\frac{\mathrm{d}P}{\rho} = V^2\left(\frac{\mathrm{d}\rho}{\rho} + \frac{\mathrm{d}A}{A}\right). \quad (2.32)$$

And by definition the speed of sound is

$$a^2 = \left(\frac{\partial P}{\partial \rho}\right)_s \approx \frac{\mathrm{d}P}{\mathrm{d}\rho}. \quad (2.33)$$

And

$$a^2\mathrm{d}\rho \approx \mathrm{d}P. \quad (2.34)$$

Substituting Eq. 2.34 into Eq. 2.32 results in

$$\frac{\mathrm{d}\rho}{\rho} = \frac{V^2}{a^2}\left(\frac{\mathrm{d}\rho}{\rho} + \frac{\mathrm{d}A}{A}\right) = Ma^2\left(\frac{\mathrm{d}\rho}{\rho} + \frac{\mathrm{d}A}{A}\right). \quad (2.35)$$

And solving Eq. 2.35 for dρ/ρ gives us

$$\frac{\mathrm{d}\rho}{\rho} = \left(\frac{Ma^2}{1 - Ma^2}\right)\frac{\mathrm{d}A}{A}. \quad (2.36)$$

Substitute Eq. 2.36 into the differential form of conservation of mass gives as

$$\left(\frac{Ma^2}{1 - Ma^2}\right)\frac{\mathrm{d}A}{A} + \frac{\mathrm{d}A}{A} + \frac{\mathrm{d}u}{u} = 0. \quad (2.37)$$

And solving for $\frac{\mathrm{d}V}{V}$ results in

$$\frac{\mathrm{d}V}{V} = -\frac{1}{1 - Ma^2}\frac{\mathrm{d}A}{A}. \quad (2.38)$$

Also, $\frac{\mathrm{d}V}{V}$ is equal to

Table 2.1 Area effects on gas dynamics

Pressure changes	$\frac{dP}{P} = \gamma Ma^2\left(\frac{1}{1-Ma^2}\right)\frac{dA}{A}$
Density changes	$\frac{d\rho}{\rho} = \left(\frac{Ma^2}{1-Ma^2}\right)\frac{dA}{A}$
Velocity changes	$\frac{dV}{V} = -\frac{1}{1-Ma^2}\frac{dA}{A}$

$$\frac{dV}{V} = \frac{-dP}{\rho V^2}. \tag{2.39}$$

Such that

$$\frac{dP}{P} = \frac{\rho V^2}{P}\frac{1}{1-Ma^2}\frac{dA}{A} = \gamma Ma^2\left(\frac{1}{1-Ma^2}\right)\frac{dA}{A}. \tag{2.40}$$

A summary table is given (see Table 2.1).

We'll now use Table 2.1 to see how pressure, density, and velocity change through a converging and then diverging nozzle. The results of this analysis are given below.

We see in Fig. 2.1 that pressure is continually decreasing which is consistent with a fluid system; we also see the density decrease in a similar fashion with the pressure, which is consistent with ideal gas behavior. Further, we see the area is decreasing for Mach < 1 and then increasing for Mach > 1 and that the velocity continually increases through the device and drastically increases around Mach 1.

The portion of the device with minimal area can be the point of a choking [1, 2]. Choking is a condition where the pressure differential between stagnation pressure and downstream pressure does not dictate the mass flux. It can be shown for $\frac{P_0}{P*} > 2$, where P_0 is the stagnation pressure and $P*$ is the pressure at the choke point, that

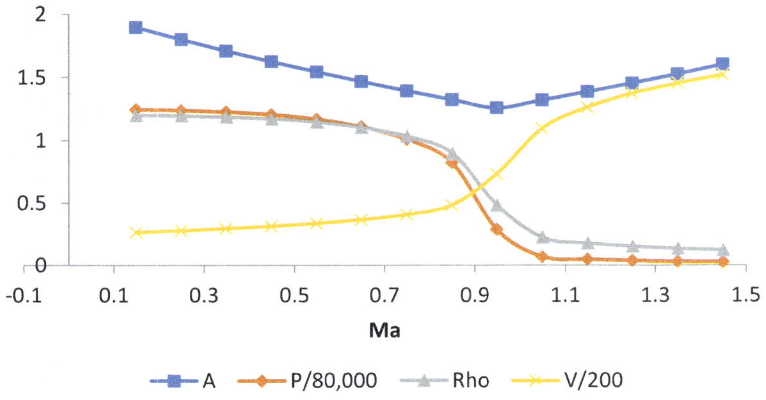

Fig. 2.1 Converging/diverging nozzle [2]

the Mach number at the choke point is always 1 and is consistent with Fig. 2.1; the critical flux through a choke point is the maximal flux allowed [1, 2] through the system.

2.7 Generalized Equations (Constant Area)

For flow through a pipe with constant cross-sectional and adiabatic conditions, a non-dimensional form of the conservation equations can be developed to answer the following types of questions [1]:

1. Determining P_1 or P_2 for a given set of conditions.
2. Determining Flow Rate for a given set of conditions.
3. Determining pipe diameter for a given set of conditions.

These three scenarios can be determined as follows:
Unknown driving force (P_2)

$$P_2 = f_1(T_1, P_1, G, D, L). \tag{2.41}$$

Unknown flow rate (G)

$$G = f_2(T_1, P_1, P_2, D, L). \tag{2.42}$$

Unknown diameter (D)

$$D = f_3(T_1, P_1, P_2, G, L). \tag{2.43}$$

Given
Conservation of mass as

$$G = \rho V = \text{constant}. \tag{2.44}$$

Conservation of energy as

$$h + \frac{1}{2}V^2 = \text{constant}. \tag{2.45}$$

Conservation of momentum as

$$\frac{\mathrm{d}P}{\rho} - V\mathrm{d}V = -2\frac{f}{D}V^2\mathrm{d}L. \tag{2.46}$$

The derivations given below assume a thermodynamically perfect gas that obeys the adiabatic relationships.

The above equations can be non-dimensionalized into the following form:

$$\frac{1}{P}\frac{\mathrm{d}P}{\mathrm{d}L} = -\gamma Ma^2 \frac{\left[1 + (\gamma - 1)Ma^2\right]}{2\left(1 - Ma^2\right)}\left[\frac{4f}{D}\right] \tag{2.47}$$

$$\frac{1}{Ma^2}\frac{\mathrm{d}Ma^2}{\mathrm{d}L} = \frac{\gamma Ma^2\left[1 + (\gamma - 1)\frac{Ma^2}{2}\right]}{1 - Ma^2}\left[\frac{4f}{D}\right] \tag{2.48}$$

$$\frac{1}{V}\frac{\mathrm{d}V}{\mathrm{d}L} = \frac{-\mathrm{d}\rho}{\rho} = \frac{\gamma Ma^2}{2\left(1 - Ma^2\right)}\left[\frac{4f}{D}\right] \tag{2.49}$$

$$\frac{1}{T}\frac{\mathrm{d}T}{\mathrm{d}L} = -\gamma \frac{(\gamma - 1)Ma^2}{2\left(1 - Ma^2\right)}\left[\frac{4f}{D}\right] \tag{2.50}$$

$$\frac{1}{P_0}\frac{\mathrm{d}P_0}{\mathrm{d}L} = -\gamma \frac{Ma^2}{1 + \gamma Ma^2}\left[\frac{4f}{D}\right] \tag{2.51}$$

$$\frac{\mathrm{d}F}{F} = \frac{\mathrm{d}P}{P} + \frac{\gamma Ma^2}{1 + \gamma Ma^2}\frac{\mathrm{d}Ma^2}{Ma^2}, \tag{2.52}$$

where

$$F = PA + \rho AV^2 = PA\left(1 + \gamma Ma^2\right). \tag{2.53}$$

When $Ma = 1$, Eq. 2.47 integrated from $(L = 0, Ma)$ to $(L = L^*, Ma = 1)$ results in

$$4\frac{fL^*}{D} = \frac{1 - Ma^2}{\gamma Ma^2} + \frac{\gamma + 1}{2\gamma}\ln\left[\frac{(\gamma + 1)Ma^2}{2 + (\gamma - 1)Ma^2}\right]. \tag{2.54}$$

And it is assumed that f is constant so that

$$4\frac{f\,\Delta L}{D} = \left(\frac{4fL^*}{D}\right)_1 - \left(\frac{4fL^*}{D}\right)_2. \tag{2.55}$$

The other equations above can also be integrated from $(L = 0, Ma)$ to $(L = L^*, Ma = 1)$ results in

$$\frac{P}{P^*} = \frac{\sqrt{f(Ma, \gamma)}}{Ma} \tag{2.56}$$

$$\frac{T}{T^*} = \left(\frac{c}{c^*}\right)^2 = f(Ma, \gamma) \tag{2.57}$$

$$\frac{\rho}{\rho^*} = \frac{V^*}{V} = \frac{1}{Ma\sqrt{f(Ma, \gamma)}}. \tag{2.58}$$

And

$$\frac{P_0}{P_0^*} = \frac{1}{Maf(Ma, \gamma)^{\gamma+1/2(\gamma-1)}}, \tag{2.59}$$

where $f(Ma, \gamma)$ equals

$$f(Ma, \gamma) = \frac{\gamma + 1}{2 + (\gamma - 1)Ma^2}. \tag{2.60}$$

Equations 2.55 through 2.60 can then be utilized to address the three scenarios outlined at the beginning of this section. It needs to be noted that in a long pipe a super-sonic flow tends to slow down to the sonic point or a sub-sonic flow tends to speed up to the sonic point and this is a useful boundary condition for solutions.

2.8 Conservation Principles for Two-Phase Flow

Some basic definition associated with two-phase flow need to be provided and are given below.

Quality

Quality (x) of a fluid is defined as

$$x = \frac{m_g}{m_l + m_g} = \frac{\dot{m}_g}{\dot{m}_l + \dot{m}_g}, \tag{2.61}$$

where m_g is the mass of gas and m_l is the mass of liquid.

Void Ratio

Void fraction (α) is defined as

$$\alpha = \frac{V_g}{V_l + V_g} = \frac{Q_g}{Q_l + Q_g}, \tag{2.62}$$

where V_g is the volume of gas, V_l is the volume of liquid, Q_g is the volumetric discharge of gas, and Q_l is the volumetric discharge of liquid.

Mixture Density

The density of a mixture (ρ_{mix}) is defined as

$$\rho_{\mathrm{mix}} = \alpha\rho_{\mathrm{g}} + (1-\alpha)\rho_{\mathrm{l}}, \tag{2.63}$$

where ρ_{mix} is the density of the mixture, ρ_{g} is the density of gas, and ρ_{l} is the density of liquid.

Slip

There are two-phase flows where the liquid portion and gas portion of the flow are separated; the liquid flows in the core with a gas sheath. When the flow is separated, the liquid and gas portions of the flow move at difference velocities and slip accounts for this difference. Slip (k) is defined as

$$k = \frac{v_{\mathrm{g}}}{v_{\mathrm{l}}} = \frac{\rho_{\mathrm{l}}}{\rho_{\mathrm{g}}}\frac{x(1-\alpha)}{\alpha(1-x)}, \tag{2.64}$$

where v_{g} is the velocity of gas and v_{l} is the velocity of liquid.

One particular model to estimate slip is given as

$$k \approx \left(\frac{\rho_{\mathrm{l}}}{\rho_{\mathrm{g}}}\right)^{3}, \tag{2.65}$$

which will be derived in Chap. 8.

For a steady, uniform flow and from Newton's Second Law

$$(+) \rightarrow \sum \mathrm{d}F_x = \mathrm{d}\left[\dot{m}_{\mathrm{g}}V_{\mathrm{g}} + \dot{m}_{\mathrm{l}}V_{\mathrm{l}}\right] = \mathrm{d}F_{\mathrm{xp}} + \mathrm{d}F_{\mathrm{xg}} + \mathrm{d}F_{\mathrm{xw}} = 0, \tag{2.66}$$

or

$$= -A_x\mathrm{d}P - \left[\rho_{\mathrm{L}}(1-\epsilon) + \rho_{\mathrm{g}}\epsilon\right]A_x\mathrm{d}z - \left[\tau_{\mathrm{w,l}} + \tau_{\mathrm{w,g}}\right]W_{\mathrm{p}}\mathrm{d}x. \tag{2.67}$$

Dividing by A_x * dx and solving for − dP/dx results in

And

$$-\frac{\mathrm{d}P}{\mathrm{d}x} = \left[\rho_{\mathrm{L}}(1-\epsilon) + \rho_{\mathrm{g}}\epsilon\right]g\frac{\mathrm{d}z}{\mathrm{d}x} + \left[\tau_{\mathrm{w,l}} + \tau_{\mathrm{w,g}}\right]\left(\frac{4}{D_{\mathrm{h}}}\right)$$
$$+ \frac{1}{A_x}\frac{\mathrm{d}}{\mathrm{d}x}\left[\dot{m}_{\mathrm{g}}V_{\mathrm{g}} + \dot{m}_{\mathrm{l}}V_{\mathrm{l}}\right]\left[\dot{m}_{\mathrm{g}}V_{\mathrm{g}} + \dot{m}_{\mathrm{l}}V_{\mathrm{l}}\right] \tag{2.68}$$

or

$$\frac{-\mathrm{d}P}{\mathrm{d}x} = -\left(\frac{\mathrm{d}P}{\mathrm{d}x}\right)_{\mathrm{f}} - \left(\frac{\mathrm{d}P}{\mathrm{d}x}\right)_{\mathrm{g}} - \left(\frac{\mathrm{d}P}{\mathrm{d}x}\right)_{\mathrm{acc}}. \tag{2.59}$$

It has been shown that

$$-\left(\frac{\mathrm{d}P}{\mathrm{d}X}\right)_{\mathrm{f}} = 4\frac{f_{\mathrm{m}}}{\rho_{\mathrm{m}}D_{\mathrm{h}}}\left(\frac{\rho_{\mathrm{m}}V_{\mathrm{m}}^2}{2}\right) = 2\frac{f_{\mathrm{m}}}{\rho_{\mathrm{m}}D_{\mathrm{h}}}G_{\mathrm{m}}^2. \tag{2.70}$$

And

$$-\left(\frac{\mathrm{d}P}{\mathrm{d}X}\right)_{\mathrm{acc}} = \rho_{\mathrm{m}}V_{\mathrm{m}}\frac{\mathrm{d}V_{\mathrm{m}}}{\mathrm{d}X} = G_{\mathrm{m}}^2\frac{\mathrm{d}v_{\mathrm{m}}}{\mathrm{d}X}, \tag{2.71}$$

where the average specific volume is defined as

$$\upsilon_{\mathrm{m}} = \upsilon_{\mathrm{g}}x + (1-x)\upsilon_{\mathrm{l}}. \tag{2.72}$$

And the derivative $\frac{\mathrm{d}\upsilon_m}{\mathrm{d}X}$ is

$$\frac{\mathrm{d}\upsilon_{\mathrm{m}}}{\mathrm{d}X} = x\frac{\mathrm{d}\upsilon_{\mathrm{g}}}{\mathrm{d}X} + \left(\upsilon_{\mathrm{g}} - \upsilon_{\mathrm{l}}\right)\frac{\mathrm{d}x}{\mathrm{d}X} = x\frac{\mathrm{d}\upsilon_{\mathrm{g}}}{\mathrm{d}P}\frac{\mathrm{d}P}{\mathrm{d}X} + \upsilon_{\mathrm{gl}}\frac{\mathrm{d}x}{\mathrm{d}X}. \tag{2.73}$$

Substitution of Eq. 2.73 into Eq. 2.71 and substitution of Eqs. 2.70 and 2.71 into Eq. 2.68 results in

$$-\frac{\mathrm{d}P}{\mathrm{d}X} = \frac{2\frac{f_{\mathrm{m}}}{\rho_{\mathrm{m}}D_{\mathrm{h}}}G_{\mathrm{m}}^2 + G_{\mathrm{m}}^2\upsilon_{\mathrm{gl}}\frac{\mathrm{d}x}{\mathrm{d}X} + \rho_{\mathrm{m}}g\frac{\mathrm{d}z}{\mathrm{d}X}}{1 + G_{\mathrm{m}}^2 x\frac{\mathrm{d}\upsilon_{\mathrm{g}}}{\mathrm{d}P}}. \tag{2.74}$$

There are a number of simplifications and particular solutions to Eq. 2.74 [1]

1. High pressure and moderate pressure gradient, Eq. 2.74 reduces to

$$-\frac{\mathrm{d}P}{\mathrm{d}X} = 2\frac{f_{\mathrm{m}}}{\rho_{\mathrm{m}}D_{\mathrm{h}}}G_{\mathrm{m}}^2 + G_{\mathrm{m}}^2\upsilon_{\mathrm{gl}}\frac{\mathrm{d}x}{\mathrm{d}X} + \rho_{\mathrm{m}}g\frac{\mathrm{d}z}{\mathrm{d}X}. \tag{2.75}$$

2. For frozen flow ($x_1 = x_2$), the second terms in the numerator of Eq. 2.74 are

$$G_{\mathrm{m}}^2\upsilon_{\mathrm{gl}}\frac{\mathrm{d}x}{\mathrm{d}X} \approx 0. \tag{2.76}$$

3. For flashing flow (see Eq. 2.74),

$$x\frac{\mathrm{d}\upsilon_{\mathrm{g}}}{\mathrm{d}P}\frac{\mathrm{d}P}{\mathrm{d}X} \approx 0. \tag{2.77}$$

Such that

$$\frac{\mathrm{d}\upsilon_{\mathrm{m}}}{\mathrm{d}X} \approx \upsilon_{\mathrm{gl}}\frac{\mathrm{d}z}{\mathrm{d}X}. \tag{2.78}$$

4. For non-flashing flow and an ideal mixture, the maximum mass flux is

$$G_{\mathrm{m}}^{*} = c_{\mathrm{m}}\rho_{\mathrm{m}} = \rho_{\mathrm{m}}\sqrt{\gamma\left(\frac{\partial P}{\partial \rho_{\mathrm{m}}}\right)_{\mathrm{T}}} = \sqrt{\frac{\rho_{\mathrm{m}}\gamma P}{\epsilon}}. \tag{2.79}$$

References

1. R. Darby, *Chemical Engineering Fluid Mechanics*, 2nd edn. (Marcel Dekker, New York, NY, 2001)
2. H. Foust, *Thermodynamics, Gas Dynamics, and Combustion* (Springer, New York, NY, 2022)

Chapter 3
Gaseous Speed of Sound, Single Phase (Component)

Abstract In this chapter, the definition for speed of sound through a gas is given and then a simple equation is developed for an ideal, perfect gas. Cubic equations of state to include van der Waal, Redlich–Kwong, and Redlich–Kwong–Soave are derived as well as the relationship between speed of sound (isothermal) and speed of sound (isentropic). The chapter goes on to provide speed of sound equations for a van der Waal and Redlich–Kwong gas and shows the effects of assuming an ideal, perfect gas on speed of sound when the gas is allowed to be cubic with a temperature varying specific heat.

Keywords Speed of sound through an ideal gas · Cubic equations of state · Real gas effects on speed of sound · Varying specific heat · Speed of sound through cubic equation of state gases

3.1 Preface

In this chapter, the speed of sound for an ideal, perfect gas will be derived from first principles; the chapter will then present various equations of state particular to a gas; the chapter will go on to speed of sound for two cubic equation of state gases (van der Waal and Redlich–Kwong); and delve into the effects of temperature varying specific heat and equation of state on speed of sound through air.

3.2 Speed of Sound, Ideal, Perfect Gas

Figure 3.1 provides the conditions for a substance moving at the speed of sound.

From conservation of mass,

$$\rho A c = (\rho + \Delta\rho) A (c - \Delta V). \tag{3.1}$$

H. Foust, *Speed of Sound and Critical Flow*, Shock Wave and High Pressure Phenomena, https://doi.org/10.1007/978-981-97-5918-7_3

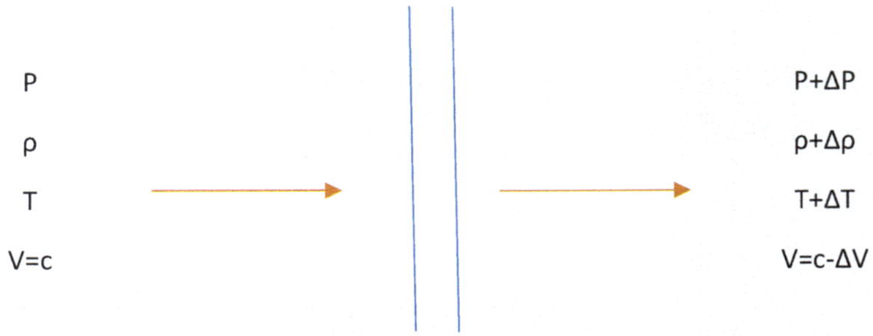

Fig. 3.1 Control volume for gas moving at speed of sound [1, 2]

Expanding becomes

$$\rho Ac = \rho Ac + \Delta\rho Ac - \rho A\Delta V - \Delta\rho A\Delta V, \tag{3.2}$$

which simplifies to

$$\rho\Delta V + \Delta\rho\Delta V = \Delta\rho c. \tag{3.3}$$

And solving for ΔV

$$\Delta V = \frac{c\Delta\rho}{\rho + \Delta\rho}. \tag{3.4}$$

From conservation of momentum,

$$PA - (P + \Delta P)A = \dot{m}(V_{\text{out}} - V_{\text{in}}) = \rho Ac(c - \Delta V - c). \tag{3.5}$$

Divide by A and simplify

$$\Delta P = \rho c\Delta V. \tag{3.6}$$

Substituting Eq. 3.4 for ΔV results in

$$\Delta P = \rho\frac{c^2\Delta\rho}{\rho + \Delta\rho}. \tag{3.7}$$

And solving for c^2

$$c^2 = \frac{\Delta P}{\Delta\rho}\left(\frac{\rho + \Delta\rho}{\rho}\right) = \frac{\Delta P}{\Delta\rho}\left(1 + \frac{\Delta\rho}{\rho}\right). \tag{3.8}$$

Taking the limit as $\Delta\rho$ goes to zero results in (where it can be shown that ΔP can be shown to be quite some and consequently $\Delta\rho$ as well)

$$c^2 = \lim_{\Delta\rho\to 0} \frac{\Delta P}{\Delta\rho}\left(1 + \frac{\Delta\rho}{\rho}\right)_s = \left(\frac{\partial P}{\partial\rho}\right)_s, \tag{3.9}$$

where the subscript implies along an isentropic path.

Equation 3.9 is the definition of speed of sound.

3.2.1 Speed of Sound in an Ideal Gas

Entropy is a state function defined for a reversible process as

$$\mathrm{d}s = \frac{\delta q}{T}. \tag{3.10}$$

And

$$\mathrm{d}u = \delta q - \delta w. \tag{3.11}$$

Substitution of Eq. 3.11 into 3.10 and applying the definition of δw result in

$$T\mathrm{d}s = \mathrm{d}u + P\mathrm{d}\upsilon. \tag{3.12}$$

If $\nu = 1/\rho$, then $\mathrm{d}\nu = -\frac{1}{\rho^2}\mathrm{d}\rho$. Thus

$$T\mathrm{d}s = \mathrm{d}u - \frac{P}{\rho^2}\mathrm{d}\rho \tag{3.13}$$

u is a function of both P and ρ such that

$$\mathrm{d}u = \frac{\partial u}{\partial P}_\rho \mathrm{d}P + \frac{\partial u}{\partial\rho}_P \mathrm{d}\rho. \tag{3.14}$$

Substitution of Eq. 3.14 into 3.13 results in

$$T\mathrm{d}s = \frac{\partial u}{\partial P}_\rho \mathrm{d}P + \frac{\partial u}{\partial\rho}_P \mathrm{d}\rho - \frac{P}{\rho^2}\mathrm{d}\rho. \tag{3.15}$$

When $\mathrm{d}s$ equals zero then

$$\frac{\partial P}{\partial\rho}_{\mathrm{d}s=0} = \frac{-\frac{\partial u}{\partial\rho}_P + \frac{P}{\rho^2}}{\frac{\partial u}{\partial P}_\rho}. \tag{3.16}$$

For an ideal, perfect gas

$$u = c_{\mathrm{v}}T + \text{constant}. \tag{3.17}$$

And

$$u = \frac{1}{\gamma - 1}\frac{P}{\rho} + \text{constant}, \tag{3.18}$$

where the relationship $c_{\mathrm{p}} - c_{\mathrm{v}} = R$ has been utilized.

Derivatives for $\frac{\partial u}{\partial P}_{\rho}$, $\frac{\partial u}{\partial \rho}_{P}$ and $\frac{\partial P}{\partial \rho}_{s}$ are given below:

$$\frac{\partial u}{\partial P}_{\rho} = \frac{1}{\gamma - 1}\frac{1}{\rho} \tag{3.19}$$

$$\frac{\partial u}{\partial \rho}_{P} = \frac{-1}{\gamma - 1}\frac{P}{\rho^2}. \tag{3.20}$$

And

$$\frac{\partial P}{\partial \rho}_{s} = \frac{\frac{1}{\gamma-1}\frac{P}{\rho^2} + \frac{P}{\rho^2}}{\frac{1}{\gamma-1}\frac{1}{\rho}} = \frac{P}{\rho} + (\gamma - 1)\frac{P}{\rho} = \gamma\frac{P}{\rho} = \gamma RT, \tag{3.21}$$

which is the definition for the speed of sound for an ideal, perfect gas.

Naturally, if the conditions dictate another equation of state, then the equation for speed of sound will differ from Eq. 3.21. For a van der Waal equation of state or Redlich–Kwong equation of state, the speed of sound is derived below.

3.3 Equations of State

An equation of state simply relates the states temperature, pressure, and density. More advanced models that relate pressure to specific volume are typically of the form:

$$P = v^3 + av^2 + bv + c, \tag{3.22}$$

where the values of the coefficients are based on observations made from experimental and theoretical considerations, which will be discussed further below.

The simplest form of Eq. 3.22 is

$$P = \frac{a}{v}, \tag{3.23}$$

where a is equal to

$$a = RT. \tag{3.24}$$

This is the ideal gas model and will be discussed next.

3.3.1 Ideal Gas Behavior

It has been found that as long as the density isn't too great, then gas behaviors in the following manner

$$PV = nR_\mathrm{u}T, \tag{3.25}$$

where R_u is the universal gas constant.

It has also been found that the kinetic theory of gases [1] explains well why the equation given as Eq. 3.25 tends to work, the form it takes, and when it fails to work. It has been found that if the density increases enough, then the gas is no longer considered ideal and a correction is made that is given as

$$PV = nZR_\mathrm{u}T, \tag{3.26}$$

where Z is the compressibility factor and non-ideal behavior occurs when $Z \neq 1$ and Z is of the functional form

$$Z = f(P_\mathrm{r}, T_\mathrm{r}, \omega), \tag{3.27}$$

where P_r is the reduced pressure, T_r is the reduced temperature, and ω is a parameter to account for the amount of deviation from spherical behavior of the molecule [1] and will be discussed further below in a separate section.

Where T_r is defined as

$$T_\mathrm{r} = \frac{T}{T_\mathrm{c}}. \tag{3.28}$$

And P_r is defined as

$$P_\mathrm{r} = \frac{P}{P_\mathrm{c}}. \tag{3.29}$$

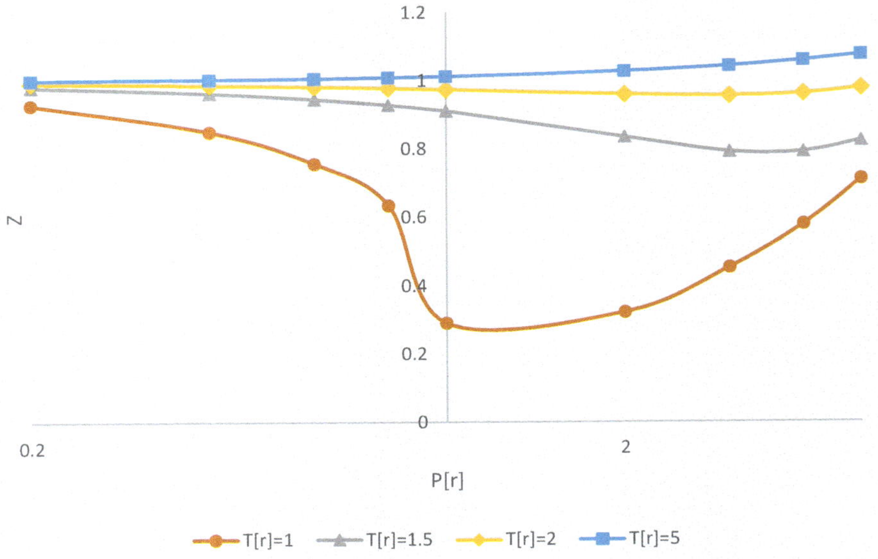

Fig. 3.2 Lee Kessler chart [1]

3.3.2 Lee/Kessler Charts

In the previous section, it was alluded to that when the density of the gas increases enough, then the gas no longer acts ideal. If we define the ideal gas equation in terms of density

$$\rho = \frac{P}{RT}. \tag{3.30}$$

So that for a fixed temperature, increasing pressure will increase density and for a fixed pressure, decreasing temperature will increase density.

Several pure substances where sphericity of the molecule isn't relevant for graphed on a graph of Z versus P_r for various T_r (see Fig. 3.2).

It can be shown from this figure that the two observations of about density increasing for an ideal gas hold as well in the figure.

3.3.3 Deviations from Spherical Charge (Acentric Factor)

We did see that the Lee Kessler Chart provides a means to determine the deviation from ideal behavior and is of the form

$$Z = f(T_r, P_r), \tag{3.31}$$

where T_r is reduced temperature, and P_r is reduced pressure

The relationship given as Eq. 3.31 works well for simple fluids, but it has been shown that for more complicated fluids another factor is required to accurately predict Z. This factor is known as the acentric factor.

It can be shown from thermodynamics [1] that

$$\frac{dP}{P} = \frac{\Delta h_{Sat}}{R_g} \frac{dT}{T^2}. \tag{3.32}$$

And assuming Δh_{Sat} is constant results in the Clausius–Clapeyron equation [1]

$$\ln\left(P^{Sat}\right) = -\frac{\Delta h_{Sat}}{R_g} \frac{1}{T} + C. \tag{3.33}$$

From Eq. 3.33, a plot of $\log\left(P_r^{Sat}\right)$ versus $1/T_r$ will result in a straight line and for simple fluids (argon, krypton, and xenon) all data falls along the same line, which has slope equal to − 2.3 and intercepts point (1, 0). More complicated substances have other slopes and intercept point (1, 0).

For a simple fluid, the plot of $\log\left(P_r^{Sat}\right)$ versus $1/T_r$ goes through the point

$$\left\{T_r, \log\left(P_r^{Sat}\right)\right\} = (0.7, -1). \tag{3.34}$$

Going back to the Lee Kessler chart, we would see the line $T_r = 0.7$ is the saturated vapor line and the acentric factor is defined as

$$\omega = -1 - \log\left(P_r^{Sat}\right)_{T_r=0.7}, \tag{3.35}$$

where $\omega = 0$ for a simple fluid.

For simple fluids, it is enough to define Z as

$$Z = f(T_r, P_r). \tag{3.36}$$

But for more complicated fluids, we need to resort to

$$Z = f(T_r, P_r, \omega). \tag{3.37}$$

3.3.4 Cubic Equations of State

Three cubic equations of state will be considered: van der Waals, Redlich–Kwong, and Redlich–Kwong–Soave [1]. Each has the following form

$$P = P_{\mathrm{rp}} + P_{\mathrm{A}}, \tag{3.38}$$

where P_{rp} is the repulsive force and given as

$$P_{\mathrm{rp}} = \frac{R_{\mathrm{u}}T}{\nu - b}. \tag{3.39}$$

And P_{A} is the attractive force and given as

$$P_{\mathrm{A}} = \frac{-\,a\alpha}{g(\nu, b)}, \tag{3.40}$$

where $a\alpha$ is a measure of the inter-molecular attractive forces and b is related to the size of the equivalent rigid sphere.

These cubic equations of state are based on the following considerations [1]:

1. Mimic the cubic behavior between P and ν.
2. Insure $\frac{\partial P}{\partial \nu} = 0$ and $\frac{\partial^2 P}{\partial \nu^2} = 0$ at the critical point.
3. The later cubic models also addressed the fact that all gases take on a constant specific volume at higher pressures.
4. Later models incorporated the effects addressed by acentric factors.

The Redlich–Kwong EOS [1] addresses Condition 3 and the Redlich–Kwong–Soave [1] addresses Conditions 3 and 4.

Van der Waal's Equation of State

One of the first attempts to address the deviations from ideal behavior seen as the density increases was the work of van der Waal [1]. The van der Waal equation of state is

$$\left(P + \frac{a}{\nu^2}\right)(\nu - b) = R_{\mathrm{u}}T \tag{3.41}$$

or

$$P = P_{\mathrm{R}} + P_{\mathrm{A}} = \frac{R_{\mathrm{u}}T}{\nu - b} - \frac{a}{\nu^2}. \tag{3.42}$$

These two parameters $\{a, b\}$ can be determine from the following conditions at the critical point

$$\frac{\partial P}{\partial \nu} = 0 \text{ and } \frac{\partial^2 P}{\partial \nu^2} = 0. \tag{3.43}$$

The values of $\{a, b\}$ determined from Eqs. 3.42 and 3.43 are

$$a = \frac{27}{64}\frac{R^2T_c^2}{P_c}. \tag{3.44}$$

And

$$b = \frac{RT_c}{8P_c}. \tag{3.45}$$

Going back to Eq. 3.22, you'll notice a cubic type of relationship between V and P, which Eqs. 3.38, 3.39, 3.40 accounts for. Another version of this equation in terms of Z is

$$Z^3 - aZ^2 + bz - c = 0, \tag{3.46}$$

where

$$a = \frac{P_R}{8T_R} + 1,\ b = \frac{27P_R}{64T_R^2},\ c = \frac{27P_R^2}{512T_R^3}. \tag{3.47}$$

It's been found for various reasons that the van der Waal equation of state is inaccurate near and above the critical point and other equations of state show higher accuracy in these regions. Van der Waal's equation of state is known as an example of a two-parameter cubic equation of state.

Redlich–Kwong Equation of State

In 1949, Redlich and Kwong [1] proposed the following model, which provided for a correction to the attractive force term within VW EOS and addresses *Condition 3* given above

$$P = \frac{R_uT}{v - b} - \frac{a\alpha}{v(v + b)}. \tag{3.48}$$

Given

$$\alpha = \frac{1}{\sqrt{T}}. \tag{3.49}$$

Using the conditions stated as Eq. 3.43, the values for $\{a, b\}$ are

$$a = 0.4748\frac{R^2T_c^{2.5}}{P_c},\ b = 0.08664\frac{RT_c}{P_c}. \tag{3.50}$$

Redlich–Kwong–Soave Equation of State

Soave in 1972 [1] incorporated into the Redlich–Kwong EOS *Conditions 3* and *4*

$$P = \frac{R_u T}{v - b} - \frac{a\alpha}{v(v + b)}, \quad (3.51)$$

where

$$\alpha = \left[1 + S\left(1 - \sqrt{T_R}\right)\right]^2, \quad S = 0.48 + 1.574\omega - 0.177\omega^2. \quad (3.52)$$

And satisfying the conditions states as Eq. 3.43, the values for $\{a, b\}$ are

$$a = \frac{0.42748 R^2 T_c^2}{P_c}, \quad b = \frac{0.08664 R T_c}{P_c}. \quad (3.53)$$

Given in Figs. 3.3, 3.4, and 3.5 are P_r versus specific volume for three values of reduced temperature (T_r) where the substance is carbon dioxide. It is seen as the value of T_r increases the deviation from ideal behavior decreases, which is consistent with what was stated above.

For the three cubic equations of state discussed,

$$Z^3 + \alpha Z^2 + \beta Z + \gamma = 0, \quad (3.54)$$

where the coefficients are defined in Table 3.1.

And $\{a, b\}$ are the coefficients particular to a given equations of state.

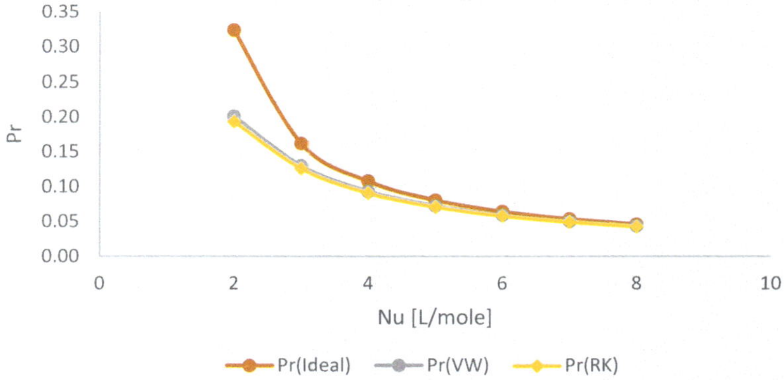

Fig. 3.3 P_r versus Nu ($T_r = 4.12$)

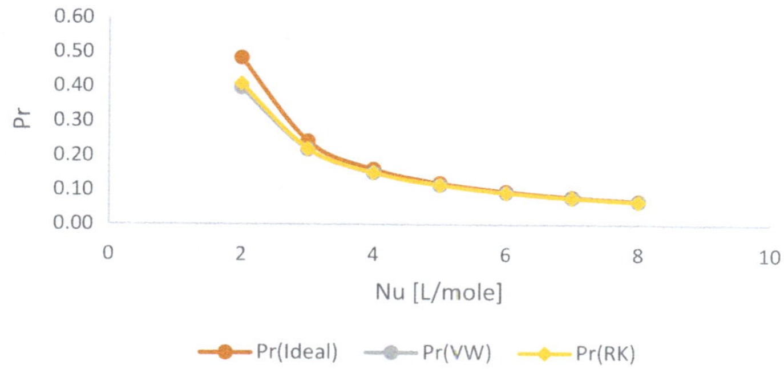

Fig. 3.4 P_r versus Nu ($T_r = 6.17$)

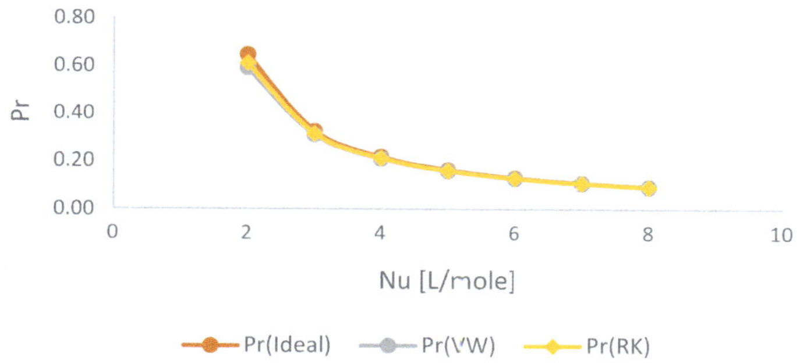

Fig. 3.5 P_r versus Nu ($T_r = 8.23$)

Table 3.1 Coefficients for compressibility factor (Z) [1]

	VW	RK	SRK
α	$-1-B$	-1	-1
β	A	$A-B-B^2$	$A-B-B^2$
γ	$-AB$	$-AB$	$-AB$
A	$\frac{aP}{(R_u T)^{2.5}}$	$\frac{aP}{R_u^2 T^{2.5}}$	$\frac{aP}{(R_u T)^{2.5}}$
B	$\frac{bP}{R_u T}$	$\frac{bP}{R_u T}$	$\frac{bP}{R_u T}$
Z	$\frac{Pv}{R_u T}$	$\frac{Pv}{R_u T}$	$\frac{Pv}{R_u T}$

3.4 Speed of Sound for Cubic Equations of State

Given the definitions for specific heat (constant volume) and specific heat (constant pressure)

$$c_v = \left(\frac{\partial u}{\partial T}\right)_v. \tag{3.55}$$

And

$$c_{\mathrm{p}} = \left(\frac{\partial h}{\partial T}\right)_{\mathrm{p}}. \tag{3.56}$$

The definition of entropy is given as

$$T\mathrm{d}s = \delta q. \tag{3.57}$$

And definition of work is given as

$$\delta w = P\mathrm{d}v. \tag{3.58}$$

Substituting Eqs. 3.57 and 3.58 into an energy balance in terms of internal energy gives us

$$\mathrm{d}u = \delta q - \delta w = T\mathrm{d}s - P\mathrm{d}v \tag{3.59}$$

or for an energy balance in terms of enthalpy

$$h = u + Pv. \tag{3.60}$$

And dh equals

$$\mathrm{d}h = \mathrm{d}u + P\mathrm{d}v + v\mathrm{d}P = T\mathrm{d}s - P\mathrm{d}v + P\mathrm{d}v + v\mathrm{d}P = T\mathrm{d}s + v\mathrm{d}P. \tag{3.61}$$

And along a constant volume path (using Eq. 3.59)

$$\left(\frac{\partial u}{\mathrm{d}T}\right)_{\mathrm{v}} = T\left(\frac{\partial s}{\partial T}\right)_{v} \tag{3.62}$$

or

$$c_{\mathrm{v}} = T\left(\frac{\partial s}{\partial T}\right)_{v}. \tag{3.63}$$

And along a constant pressure path (using Eq. 3.61)

$$\left(\frac{\partial h}{\mathrm{d}T}\right)_{\mathrm{p}} = T\left(\frac{\partial s}{\partial T}\right)_{\mathrm{p}} \tag{3.64}$$

Or

$$c_{\mathrm{p}} = T\left(\frac{\partial s}{\partial T}\right)_{\mathrm{p}}. \tag{3.65}$$

And γ can be defined as

$$\gamma = \frac{c_\mathrm{p}}{c_\mathrm{v}} = \frac{T\left(\frac{\partial s}{\partial T}\right)_\mathrm{p}}{T\left(\frac{\partial s}{\partial T}\right)_\nu}. \tag{3.66}$$

From the Maxwell relationships [1],

$$\left(\frac{\partial s}{\partial T}\right)_\mathrm{p} = -\left(\frac{\partial s}{\partial P}\right)_\mathrm{T}\left(\frac{\partial P}{\partial T}\right)_\mathrm{s}. \tag{3.67}$$

And

$$\left(\frac{\partial s}{\partial T}\right)_\nu = -\left(\frac{\partial T}{\partial \nu}\right)_\mathrm{s}\left(\frac{\partial \nu}{\partial s}\right)_\mathrm{T}. \tag{3.68}$$

Substitution of Eqs. 3.67 and 3.68 into 3.66 results in

$$\begin{aligned}\gamma &= \frac{\left(\frac{\partial s}{\partial T}\right)_\mathrm{p}}{\left(\frac{\partial s}{\partial T}\right)_\nu} = \frac{\left(\frac{\partial s}{\partial P}\right)_\mathrm{T}\left(\frac{\partial P}{\partial T}\right)_\mathrm{s}}{\left(\frac{\partial \nu}{\partial T}\right)_\mathrm{s}\left(\frac{\partial s}{\partial \nu}\right)_\mathrm{T}} = \frac{\left(\frac{\partial s}{\partial P}\right)_\mathrm{T}\left(\frac{\partial \nu}{\partial s}\right)_\mathrm{T}}{\left(\frac{\partial \nu}{\partial T}\right)_\mathrm{s}\left(\frac{\partial T}{\partial P}\right)_\mathrm{s}} \\ &= \frac{\left(\frac{\partial \nu}{\partial P}\right)_\mathrm{T}}{\left(\frac{\partial \nu}{\partial P}\right)_\mathrm{s}} = \frac{\left(\frac{\partial P}{\partial \nu}\right)_\mathrm{s}}{\left(\frac{\partial P}{\partial \nu}\right)_\mathrm{T}} = \frac{c_\mathrm{s}^2}{c_\mathrm{T}^2}.\end{aligned} \tag{3.69}$$

Equation 3.69 states that Gamma is a ratio of the squares of speed of sound (isentropic) over square of speed of sound (isothermal).

Given the following representation for van der Waal's Equation of state

$$P = \frac{RT}{\nu - b} - \frac{a}{\nu^2}. \tag{3.70}$$

And that γ is defined as

$$\gamma = \frac{c_\mathrm{p}}{c_\mathrm{v}} = \frac{\left(\frac{\partial P}{\partial \rho}\right)_\mathrm{s}}{\left(\frac{\partial P}{\partial \rho}\right)_\mathrm{T}}. \tag{3.71}$$

And the definition of speed of sound (isentropic) is

$$c^2 = \left(\frac{\partial P}{\partial \rho}\right)_\mathrm{s} \tag{3.72}$$

or in terms of ν where $\nu = 1/\rho$ is

$$c^2 = -\nu^2\left(\frac{\partial P}{\partial \nu}\right)_s = -\nu^2\left(\frac{\partial P}{\partial \nu}\right)_s = -\gamma\nu^2\left(\frac{\partial P}{\partial \nu}\right)_T. \tag{3.73}$$

And $\left(\frac{\partial P}{\partial v}\right)_T$ for a van der Waal gas is

$$\left(\frac{\partial P}{\partial \nu}\right)_T = \frac{-RT}{(\nu - b)^2} + 2\frac{a}{\nu^3}. \tag{3.74}$$

Substitution of Eq. 3.74 into Eq. 3.73 results in

$$c^2 = -\gamma\nu^2\left(\frac{\partial P}{\partial \nu}\right)_T = \gamma\left(\frac{\nu^2 RT}{(\nu - b)^2} - 2\frac{a}{\nu}\right). \tag{3.75}$$

Additionally, for an ideal gas

$$c_p - c_v = R. \tag{3.76}$$

And

$$\gamma = \frac{c_p}{c_v} = \frac{c_v + R}{c_v} = 1 + \frac{R}{c_v}. \tag{3.77}$$

Substitution of Eq. 3.77 into 3.75 results in

$$c^2 = \left(1 + \frac{R}{c_v}\right)\left(\frac{\nu^2 RT}{(\nu - b)^2} - 2\frac{a}{\nu}\right). \tag{3.78}$$

Given that a theoretical basis for determining speed of found for cubic equations of state has been established, it is not a great deal of effort to apply this theory to the Redlich–Kwong EOS.

Given

$$P = \frac{R_u T}{\nu - b} - \frac{a\alpha}{\nu(\nu + b)} = U(\nu) + V(\nu), \tag{3.79}$$

where

$$\alpha = \frac{1}{\sqrt{T}}. \tag{3.80}$$

Further,

$$\left(\frac{\partial P}{\partial \nu}\right)_T = U_\nu(\nu) - V_\nu(\nu) = \frac{-RT}{(\nu - b)^2} + \frac{\alpha a(2\nu + b)}{\left(\nu^2 + b\nu\right)^2}. \tag{3.81}$$

And the definition of speed of sound (isothermal) is

$$c_{\mathrm{T}}^2 = -\nu^2\left(\frac{\partial P}{\partial \nu}\right)_{\mathrm{T}}. \tag{3.82}$$

Substituting Eq. 3.82 into Eq. 3.81 results in

$$c_{\mathrm{T}}^2 = -\nu^2\left(\frac{\partial P}{\partial \nu}\right)_{\mathrm{T}} = \frac{\nu^2 RT}{(\nu - b)^2} - \frac{\alpha a \nu^2(2\nu + b)}{\left(\nu^2 + b\nu\right)^2}. \tag{3.83}$$

And speed of sound (isentropic) is

$$\begin{aligned} c_{\mathrm{s}}^2 &= -\nu^2\gamma\left(\frac{\partial P}{\partial \nu}\right)_{\mathrm{T}} = \frac{\nu^2\gamma RT}{(\nu - b)^2} - \frac{\alpha a \nu^2\gamma(2\nu + b)}{\left(\nu^2 + b\nu\right)^2} \\ &= \left(1 + \frac{R}{c_{\mathrm{v}}}\right)\left[\frac{\nu^2 RT}{(\nu - b)^2} - \frac{\alpha a(2\nu + b)}{(\nu + b)^2}\right]. \end{aligned} \tag{3.84}$$

3.5 Real Gas Effects on Gaseous Speed of Sound

In this section the effects of real gases to include a model for specific heat and a cubic equation of state are compared against an ideal gas model for speed of sound in air. The results of this analysis for three temperatures (300, 600, and 1200 K) are given in Figs. 3.6 and 3.7.

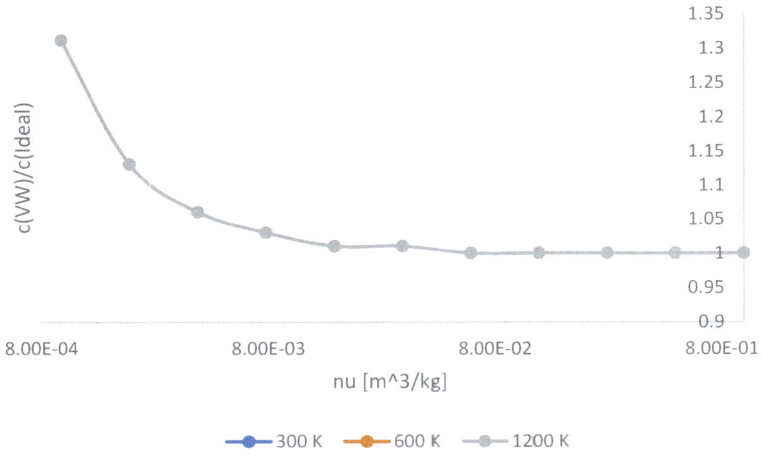

Fig. 3.6 Ratio of speed of sound versus nu (constant gamma)

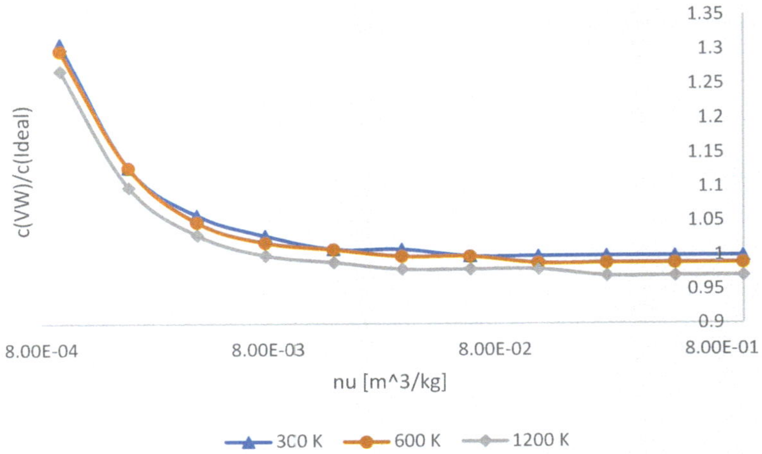

Fig. 3.7 Ratio of speed of sound versus nu (constant gamma)

The specific heat model utilized is given as

$$c_v = \frac{c_p}{\gamma} = \frac{\beta_0 + \beta_1\theta + \beta_2\theta^2 + \beta_3\theta^3}{\gamma}, \quad (3.85)$$

where $\{\beta_0, \beta_1, \beta_2, \beta_3\}$ are coefficients particular to the gas and $\theta = T\,[\mathrm{K}]/1000$.

Figure 3.6 shows graphs speed of sound (van der Waal) over speed of sound (ideal) versus specific volume where it can be seen that a deviation of 5% or greater occurs below 0.0045 m^3/kg. Both models for speed of sound utilized Eq. 3.77.

In Fig. 3.7, speed of sound (van der Waal) over speed of sound (ideal) versus specific volume is again graphed; the γ associated with the van der Waal gas is allow to vary whereas the γ associated with the ideal gas is fixed at 1.4. Again, it is seen that a deviation of 5% or greater occurs below 0.0045 m^3/kg.

This delineation of about 0.0045 m^3/kg is seen independent of temperature and whether or not the ideal gas utilized a constant value for γ.

Due to the independences of the "cut-off" value of ν around 0.0045 m^3/kg, a criterion was created for air and is given in terms of both ν_R (Eq. 3.88) and T_r/P_r (Eq. 3.89).

$$\nu < 6.5\nu_c = 3(6.5)b \approx 20b = 20\frac{RT_c}{8P_c} \rightarrow \frac{c_{vw}}{c_{ideal}} > 1 \quad (3.86)$$

$$P \approx \frac{RT}{\nu} \quad (3.87)$$

or

$$\nu \approx \frac{RT}{P} < 20\frac{RT_c}{8P_c} \tag{3.88}$$

$$\frac{T_r}{P_r} < \frac{20}{8} \tag{3.89}$$

or

$$\frac{T_r}{P_r} < 2.5 \rightarrow \frac{c_{vw}}{c_{ideal}} > 1. \tag{3.90}$$

In summary, and particular to air, below $\nu_r < 6.5$ and $\frac{T_r}{P_r} < 2.5$ an ideal gas speed of sound equation is not appropriate, and a cubic equation of state is required. Again, particular to air, the effects of varying specific heat on speed of sound appear to be minimal. Additionally, near the critical point Redlich–Kwong EOS mimics gas behavior better than the van der Waals EOS.

References

1. H. Foust, *Thermodynamics, Gas Dynamics, and Combustion* (Springer, New York, NY, 2022)
2. Speed of Sound. https://en.wikipedia.org/wiki/Speed_of_sound. Accessed 22 May 2023

Chapter 4
Speed of Sound Through Liquid/Gas Systems

Abstract This chapter begins by delineating regimes of flow for a liquid/gas and goes on to provide a discussion on speed of sound through a liquid/gas without significant surface tension, speed of sound through a liquid/gas with significant surface tension, speed of sound through a mixture of gases, and speed of sound through a liquid/gas based on the omega method.

Keywords Liquid/gas flows · Regimes of flow · Surface tension · Omega method · Karplus

4.1 Preview

We have seen for a pure gas that acts ideal the speed of sound is defined as

$$c_s^2 = \gamma R_g T. \tag{4.1}$$

We have also seen that speed of sound can be defined for other equations of state and did this for van der Waal and Redlich–Kwong gases.

In this chapter, we will develop the speed of sound through a mixture of liquid and gas [1–8]. It will be shown that the speed of sound of a mixture is below the speed of sound of either the pure liquid or the pure gas. This can be mathematically depicted as

$$c_s^2\,(\text{pure liquid}) > c_{\text{mixture}}^2. \tag{4.2}$$

H. Foust, *Speed of Sound and Critical Flow*, Shock Wave and High Pressure Phenomena, https://doi.org/10.1007/978-981-97-5918-7_4

And

$$c_s^2\,(\text{pure gas}) > c^2_{\text{mixture}}, \tag{4.3}$$

where

$$c_s^2\,(\text{pure liquid}) > c_s^2\,(\text{pure gas}). \tag{4.4}$$

The chapter will end by using a method appropriate for homogenous fluids and known as the omega method. The omega method will be more formally introduced in Chap. 9 and is the subject of the entire chapter.

4.2 Catalogue of Liquid/Gas Behavior

Given in Fig. 4.1 are the various regimes of flow for a liquid/gas system within a horizontal pipe. Each type of flow is further discussed in Table 4.1. The regime of flow is dependent on the volumetric flow rate and tends to go from bubble flow to mist flow as the volumetric discharge increases.

When we begin to ask questions about quantities such as critical flow in a liquid/gas system, we need to understand there are many types of flow that can be encountered (see Fig. 4.2), and these flows are dependent on a combination of the liquid and gas flow rates. Many of the critical flow models for two phase systems have either assumed homogenous flow or annular flow. Both of these flows are discussed below.

In **homogenous flows**, the two regimes (liquid flow and gas flow) are assumed to have some average associated with all properties such as density, pressure or flow velocity; there is equilibrium between the phases. In **annular flow**, and assuming the conduit is a pipe, the gas flows in an inner core with a higher velocity and the liquid flows as an outer sheath with a lower velocity. The reason for the differing velocities is that one pressure is applied to the whole system, but that the effective density of the gas phase is much less than the effective density of the liquid phase and so the gas phase moves at a higher velocity relative to the liquid phase. A slip velocity is created and may result in no equilibrium between the phases.

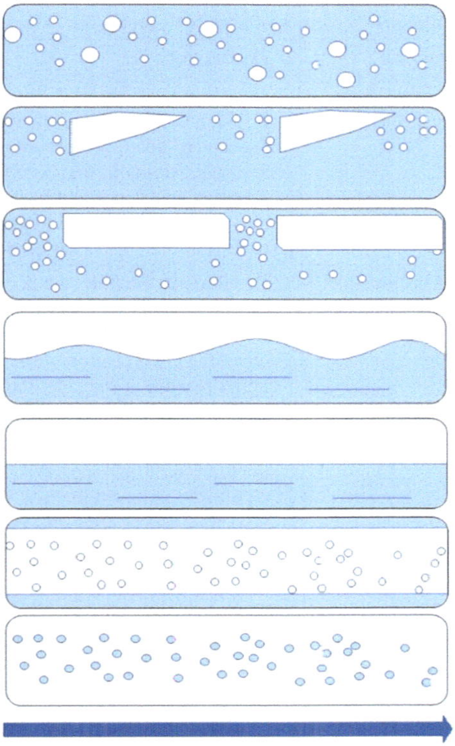

Fig. 4.1 Flow regimes in horizontal flow from top to bottom: bubble flow, plug flow, slug flow, wavy flow, stratified flow, annular flow, and mist flow for a horizontal pipe [9]

4.3 The Work of Karplus

A seminal work in the area of gas/liquid speed of sound was the work of Karplus [4]. From Wood [11], it was shown that

$$c^2 = \frac{1}{\rho\beta}, \tag{4.5}$$

where ρ is the density and β is the compressibility. Compressibility is defined as

$$\beta = -\frac{1}{V}\frac{\mathrm{d}V}{\mathrm{d}P} \tag{4.6}$$

and needs to be defined along a particular path.

Where along an isothermal path,

$$PV = \text{constant}. \tag{4.7}$$

Table 4.1 Regimes of flow for a liquid/gas system within a horizontal pipe [8]

Regime	Description
Bubble/dispersed bubble flow	Occurs at large liquid flow rates with little gas flow [21]. Bubbles of gas dispersed or suspended throughout the liquid continuous phase. Typical features of this flow are moving and deformed interfaces of bubbles in time and space domains and complex interactions between the interfaces. This flow can be categorized further into ideally separated, interacting bubble, churn turbulent, and clustered [22]. Due to the buoyancy force, bubbles tend to drift in the upper portion of the pipe [10]
Plug flow	Develops as the flow rate is increased while vapor flow is maintained at a low amount [21]. Plugs of gas in liquid phase where the velocity is assumed to be constant while "plugs," essentially "bullet-shaped" bubbles of gas that cover the cross section of the pipe flow intermittently through the pipe in the upper portion of the pipe due to buoyancy forces [10, 23]
Stratified flow	Gas and liquid flow where there is separation by an interface. This occurs when the gravity force dominates which causes stratification of the liquid at the bottom of the pipe. Most common in horizontal or slightly inclined pipelines. At low velocities, smooth interfaces occur whereas at greater velocities waves appear [24]
Wavy flow	Characterized by a gas–liquid flows in parallel streams, the interface between them is flat at low gas velocities, waves appear due to perturbations when velocity is increased. An example would be waves on the sea [25]
Slug flow	Defined by the intermittent sequence of liquid "slugs" containing disperse gas bubbles alternating with longer bubbles with greater width. Unsteady flow regime even when velocities are kept constant [26]
Annular flow	Occurs when a liquid film in gas–liquid flow covers the channel wall in an annulus shape with gas flowing in the core. The core can also contain liquid droplets; this case is known as annular-dispersed flow [27]
Mist/dispersed mist flow	Occurs at very high gas flow rates [21]. Characterized by a disperse phase being suspended in a continuous phase. In the case gas–liquid flow it occurs when liquid particles are suspended in a continuous gas phase [28]

And along an isentropic path

$$PV^{\gamma} = \text{constant}. \tag{4.8}$$

Applying the definition of compressibility (Eq. 4.6) to the isothermal condition results in

$$\beta_{\mathrm{T}} = -\frac{1}{V}\left\{-\frac{V}{P}\right\} = \frac{1}{P}. \tag{4.9}$$

And for isentropic conditions

$$\beta_{\mathrm{s}} = -\frac{1}{V}\left\{-\frac{V}{\gamma P}\right\} = \frac{1}{\gamma P}. \tag{4.10}$$

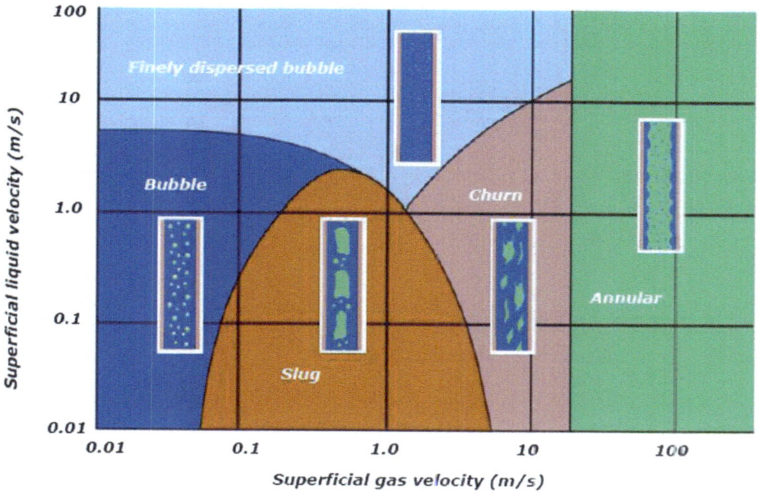

Fig. 4.2 Gas/liquid flow regimes for a vertical pipe. Reprinted from [11] licensed under the Creative Commons Attribution 4.0 International

Equation 4.5 in terms of a mixture of gas/liquid is given as

$$c^2 = \frac{1}{\rho_{\mathrm{mix}}\beta_{\mathrm{mix}}}. \tag{4.11}$$

Assuming an ideal gas, the mixture density is

$$\rho_{\mathrm{mix}} = (1-z)\rho_{\mathrm{L}} + z\rho_{\mathrm{g}}, \tag{4.12}$$

where z, volume concentration, is defined as $z = \frac{V_{\mathrm{g}}}{V_{\mathrm{l}}}$.

And the compressibility of the mixture is defined as

$$\beta_{\mathrm{mix}} = (1-z)\beta_{\mathrm{L}} + z\beta_{\mathrm{g}}. \tag{4.13}$$

Inserting Eqs. 4.12 and 4.13 into 4.11 results in

$$\frac{1}{c^2} = \left[(1-z)\rho_{\mathrm{L}} + z\rho_{\mathrm{g}}\right]\left[(1-z)\beta_{\mathrm{L}} + z\beta_{\mathrm{g}}\right] \tag{4.14}$$

or

$$\frac{1}{c^2} = z^2\rho_{\mathrm{g}}\beta_{\mathrm{g}} + (1-z)^2\rho_{\mathrm{L}}\beta_{\mathrm{L}} + z(1-z)\left(\rho_{\mathrm{g}}\beta_{\mathrm{L}} + \rho_{\mathrm{l}}\beta\beta_{\mathrm{g}}\right). \tag{4.15}$$

The speed of sound for a pure, ideal gas is

$$\frac{1}{c_g^2} = \rho_g \beta_s = \frac{\rho_g}{\gamma P}. \tag{4.16}$$

And for the liquid phase

$$\frac{1}{c_l^2} = \rho_l \beta_l. \tag{4.17}$$

The speed of sound for a mixture along an isentropic path is determined by substituting Eqs. 4.16 and 4.17 into Eq. 4.15

$$\frac{1}{c_s^2} = \frac{z^2}{c_g^2} + \frac{(1-z)^2}{c_l^2} + z(1-z)\left[\frac{\rho_l}{P\gamma} + \rho_g \beta_l\right]. \tag{4.18}$$

And along an isothermal path

$$\frac{1}{c_T^2} = \gamma \frac{z^2}{c_g^2} + \frac{(1-z)^2}{c_l^2} + z(1-z)\left[\frac{\rho_l}{P} + \rho_g \beta_l\right]. \tag{4.19}$$

The following simplifying assumptions are made

$$\rho_g \beta_l \ll \frac{\rho_l}{P}. \tag{4.20}$$

And

$$\frac{1-z}{c_l} \approx \frac{1}{c_l}. \tag{4.21}$$

Insertion of Eqs. 4.20 and 4.21 into 4.19 results in

$$\frac{1}{c_T^2} \approx \gamma \frac{z^2}{c_g^2} + \frac{1}{c_l^2} + z(1-z)\left[\frac{\rho_l}{P}\right]. \tag{4.22}$$

Equation 4.22 is valid for $1\% < z < 99\%$.

Karplus performed experimentation for a mixture of water/air bubbles ensuring that surface tension was not an issue, and the results of this experimentation are given in Fig. 4.3. Along with the data, Eqs. 4.18 and 4.19 are also portrayed. It's obvious from the figure that the system behaves isothermal.

Karplus explains this [4]. "The measurements agree well with computations if we assume that the heat produced in the compression of the bubbles is rapidly conducted into the water. Because of the very large heat capacity of the latter, the change of

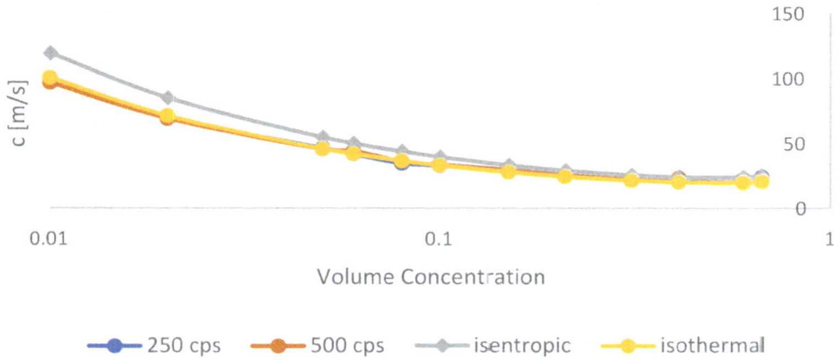

Fig. 4.3 Speed of sound versus concentration of air

temperature is quite negligible and the isothermal compressibility of the gas gives the correct results."

This theory cannot be extended to vapor–liquid systems. Karplus [4] explains. "The isothermal vibration of the bubbles shows that the results cannot be extended to a water steam bubble system since a steam bubble cannot undergo isothermal pressure changes in the presence of the liquid phase, for in such a system pressure and temperature are uniquely related. Increasing pressure will cause some of the steam to condense and decreasing pressure will cause evaporation of the water. The effect will take place even in wet steam, i.e., when the vapor is the continuous phase, and the liquid phase is dispersed in discrete droplets. In such a system small pressure changes will produce large density changes and, hence, low sound velocities may be expected."

In the next section two sections, we'll see how a vapor phase can be incorporated for models of speed of sound through liquid/gas mixtures.

4.4 Liquid/Gas System with Negligible Surface Tension

It will be shown that when the radius of gas bubbles within a liquid are larger than 10^{-4} cm, then the pressure of the gas phase (P_L) and the pressure of the liquid phase (P_G) are the same and we represent this as

$$P_L = P_G = P. \tag{4.23}$$

The density of the liquid phase is represented [5, 6] as

$$\rho_L = \rho_{LA} \exp\left[\frac{P - P_A}{K}\right], \tag{4.24}$$

where K is the bulk modulus and ρ_{LA} is the density at a reference pressure.

And the density of the gas phase is

$$\rho_G = \left(\frac{P}{c}\right)^{1/n}, \tag{4.25}$$

where P is the final pressure and c is a constant associated with polytropic expansions.

The polytropic exponent (n) of an ideal gas obeys the following relationship:

$$P_1 v_1^n = P_2 v_2^n = c, \tag{4.26}$$

where c is determined in the following way

$$PV = mR_g T, \quad Pv = R_g T. \tag{4.27}$$

Multiplying both sides by v^{n-1} results in

$$Pv^n = R_g T v^{n-1} = R_g \frac{T}{\rho^{n-1}} = c. \tag{4.28}$$

From Eq. 4.26

$$P_2 \left(\frac{1}{\rho_2}\right)^n = c \tag{4.29}$$

or

$$\rho_2 = \left(\frac{P_2}{c}\right)^{1/n}. \tag{4.30}$$

The volume can be portioned between a liquid phase and a gas phase

$$V_{Total} = V_G + V_L \tag{4.31}$$

or

$$\frac{m_L + m_G}{\rho} = \frac{m_G}{\rho_G} + \frac{m_L}{\rho_L} \tag{4.32}$$

or

$$\frac{1 + \frac{m_G}{m_L}}{\rho} = \frac{\frac{m_G}{m_L}}{\rho_G} + \frac{1}{\rho_L} \tag{4.33}$$

or

$$\frac{1+\eta}{\rho} = \frac{\eta}{\rho_G} + \frac{1}{\rho_L}, \tag{4.34}$$

where $\eta = \frac{m_G}{m_L}$ and is known as gas mass fraction.

Mass fraction of gas (η) can be related to another relationship which is the void fraction (x) through the following

$$x = \frac{V_G}{V_G + V_L} = \frac{\frac{m_G}{\rho_G}}{\frac{m_G}{\rho_G} + \frac{m_L}{\rho_L}} = \frac{m_G}{m_G + m_L\frac{\rho_G}{\rho_L}} = \frac{1}{1+\eta\frac{\rho_G}{\rho_L}}. \tag{4.35}$$

Substitution of Eqs. 4.24 and 4.25 into Eq. 4.34 results in

$$\rho = (1+\eta)\rho_{LA}\left\{\frac{\eta\rho_{LA}c^{1/n}}{P^{1/n}} + \exp\left[\frac{P_A - P}{K}\right]\right\}^{-1}. \tag{4.36}$$

Given

$$\begin{aligned}\rho &= (1+\eta)\rho_{LA}\left\{\frac{\eta\rho_{LA}c^{1/n}}{P^{1/n}} + \exp\left[\frac{P_A - P}{K}\right]\right\}^{-1} \\ &= \frac{(1+\eta)\rho_{LA}}{\frac{\eta\rho_{LA}c^{1/n}}{P^{1/n}} + \exp\left[\frac{P_A-P}{K}\right]} = \frac{A}{BP^{-1/n} + \exp\left[\frac{P_A-P}{K}\right]}\end{aligned}. \tag{4.37}$$

And

$$\frac{\partial\rho}{\partial P} = \frac{-A}{\left\{BP^{-1/n} + \exp\left[\frac{P_A-P}{K}\right]\right\}^2}\left\{\frac{-B}{n}P^{-\frac{n+1}{n}} - \frac{1}{K}\exp\left[\frac{P_A - P}{K}\right]\right\} \tag{4.38}$$

or

$$\frac{\partial\rho}{\partial P} = \frac{A}{\left\{BP^{-1/n} + C\right\}^2}\left\{\frac{B}{\gamma}P^{-\frac{n+1}{n}} - \frac{C}{K}\right\}. \tag{4.39}$$

The speed of sound can be defined for a liquid/gas system as

$$c^2 = \frac{\partial P}{\partial\rho} = \frac{\left\{BP^{-1/n} + C\right\}^2}{A\left\{\frac{B}{\gamma}P^{-\frac{n+1}{n}} + \frac{C}{K}\right\}}, \tag{4.40}$$

where

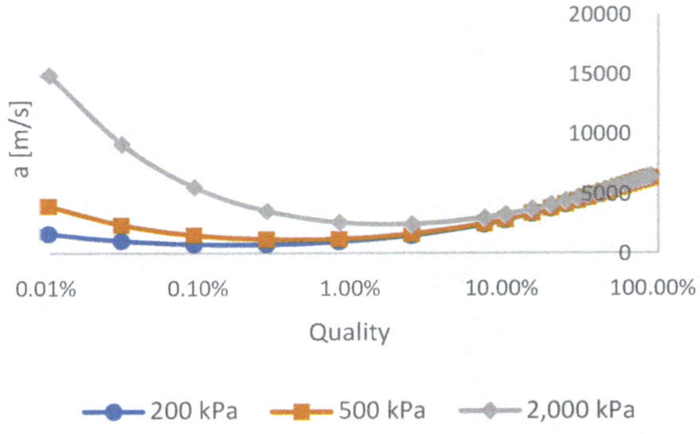

Fig. 4.4 Sonic velocity versus quality at three pressures

$$A = (1+\eta)\rho_{\mathrm{LA}},\ B = \eta\rho_{\mathrm{LA}}c^{1/n},\ C = \exp\left[\frac{P_{\mathrm{A}} - P}{K}\right] \tag{4.41}$$

or

$$c^2 = \frac{\partial P}{\partial \rho} = \frac{\left\{\eta\rho_{\mathrm{LA}}c^{1/n}P^{-1/n} + \exp\left[\frac{P_{\mathrm{A}}-P}{K}\right]\right\}^2}{(1+\eta)\rho_{\mathrm{LA}}\left\{\frac{\eta\rho_{\mathrm{LA}}c^{1/n}}{\gamma}P^{-\frac{n+1}{n}} + \frac{1}{K}\exp\left[\frac{P_{\mathrm{A}}-P}{K}\right]\right\}}. \tag{4.42}$$

And when $n = 1$, which represents an isothermal flow

$$c^2 = \frac{\partial P}{\partial \rho} = \frac{\left\{\eta\rho_{\mathrm{LA}}\left(\frac{c}{P}\right) + \exp\left[\frac{P_{\mathrm{A}}-P}{K}\right]\right\}^2}{(1+\eta)\rho_{\mathrm{LA}}\left\{\frac{\eta\rho_{\mathrm{LA}}c}{P^2} + \frac{1}{K}\exp\left[\frac{P_{\mathrm{A}}-P}{K}\right]\right\}} = \frac{[D+C]^2}{A\left[B' + \frac{1}{K}C\right]}, \tag{4.43}$$

where $D = \eta\rho_{\mathrm{LA}}\left(\frac{c}{P}\right)$, $B' = \frac{\eta\rho_{\mathrm{LA}}c}{P^2}$.

It is seen in Fig. 4.4 that with an increase in pressure there is an increase in speed of sound when surface tension is negligible.

4.5 Liquid/Gas System with Appreciable Surface Tension

When surface tension becomes appreciable, then the pressure associated with the gas bubbles (P_{G}) is no longer the same as the pressure associated with the liquid phase (P_{L}) and P_{G} is defined as

$$P_{\mathrm{G}} = P_{\mathrm{L}} + 2\frac{\sigma}{r}, \tag{4.44}$$

where σ is the surface tension and r is the radius of the bubble [3, 4, 10].

And the gas density is defined as

$$\rho_G = \frac{P_G}{c} = \frac{P_L + 2\frac{\sigma}{r}}{c}, \tag{4.45}$$

where now the path is assumed as isothermal.

And the density of the liquid is

$$\rho_L = \rho_{LA} \exp\left[\frac{P_L - P_A}{K}\right], \tag{4.46}$$

where K is the bulk modulus.

Utilizing

$$\frac{1+\eta}{\rho} = \frac{\eta}{\rho_G} + \frac{1}{\rho_L} \tag{4.47}$$

results in a mixture density of

$$\rho = \frac{1+\eta}{\frac{\eta}{\rho_G} + \frac{1}{\rho_L}} = \frac{1+\eta}{\frac{\eta c}{P_L + 2\frac{\sigma}{r}} + \frac{1}{\rho_{LA}} \exp\left(\frac{P_A - P_L}{K}\right)} \tag{4.48}$$

or in terms of P_L

$$\rho = \frac{A}{\frac{B}{P_L + C} + D \exp\left(\frac{P_A - P_L}{K}\right)}, \tag{4.49}$$

where

$$A = 1 + \eta, \; B = \eta c, \; C = 2\frac{\sigma}{r}, \; D = \frac{1}{\rho_{LA}}. \tag{4.50}$$

We can solve Eq. 4.49 for $\frac{r}{2\sigma}$ and get to

$$\frac{r}{2\sigma} = \left[\frac{\eta c}{\frac{1+\eta}{\rho} - \frac{1}{\rho_{LA}} \exp\left(\frac{P_A - P_L}{K}\right)} - P_L\right]^{-1} = \frac{1}{\beta}. \tag{4.51}$$

Further, the mass of a bubble is given as

$$m = \frac{4}{3}\pi r^3 \rho_g = \frac{4}{3}\pi r^3 \frac{P_L + 2\frac{\sigma}{r}}{c}. \tag{4.52}$$

And solving Eq. 4.51 for P_L results in

$$P_{\mathrm{L}} = \frac{3mc}{4\pi r^3} - 2\frac{\sigma}{r}. \tag{4.53}$$

And substituting Eq. 4.51 for r results in

$$P_{\mathrm{L}} = \frac{3mc}{4\pi}\left(\frac{\beta}{2\sigma}\right)^3 - \beta. \tag{4.54}$$

Equation 4.54 can equally be represented as

$$P_{\mathrm{L}} = \frac{3mc}{4\pi}\left(\frac{\beta}{2\sigma}\right)^3 - \frac{\eta c}{\frac{1+\eta}{\rho} - \frac{1}{\rho_{\mathrm{LA}}}\exp\left(\frac{P_{\mathrm{A}}-P_{\mathrm{L}}}{K}\right)} + P_{\mathrm{L}} \tag{4.55}$$

or

$$\beta^3 - \frac{32\pi\eta\sigma^3}{3m}\frac{1}{\frac{1+\eta}{\rho} - \frac{1}{\rho_{\mathrm{LA}}}\exp\left(\frac{P_{\mathrm{A}}-P_{\mathrm{L}}}{K}\right)} = 0. \tag{4.56}$$

And

$$f(\rho, P_{\mathrm{L}}) = \beta^3 - \frac{\delta}{\xi} = 0, \tag{4.57}$$

where δ is defined as

$$\delta = \frac{32\pi\eta\sigma^3}{3m}. \tag{4.58}$$

And ξ is defined as

$$\xi = \frac{1+\eta}{\rho} - \frac{1}{\rho_{\mathrm{LA}}}\exp\left(\frac{P_{\mathrm{A}}-P_{\mathrm{L}}}{K}\right) = \frac{A}{\rho} - D\exp\left(\frac{P_{\mathrm{A}}-P_{\mathrm{L}}}{K}\right). \tag{4.59}$$

Additionally,

$$\beta = \frac{\eta c}{\xi} - P_{\mathrm{L}}. \tag{4.60}$$

Given that

$$\frac{\partial f}{\partial P_{\mathrm{L}}} = 3\beta^2\frac{\partial \beta}{\partial P_{\mathrm{L}}} - \frac{\delta}{\xi^2}\frac{\partial \xi}{\partial P_{\mathrm{L}}} = 0, \tag{4.61}$$

where

$$\frac{\partial \beta}{\partial P_{\mathrm{L}}} = -\frac{\eta c}{\xi^2}\frac{\partial \xi}{\partial P_{\mathrm{L}}} - 1, \quad \frac{\partial \xi}{\partial P_{\mathrm{L}}} = \frac{1}{\rho_{\mathrm{LA}} K}\exp\left(\frac{P_{\mathrm{A}} - P_{\mathrm{L}}}{K}\right). \tag{4.62}$$

And

$$\frac{\partial f}{\partial \rho} = 3\beta^2\frac{\partial \beta}{\partial \rho} - \frac{\delta}{\xi^2}\frac{\partial \xi}{\partial \rho} = 0, \tag{4.63}$$

where

$$\frac{\partial \beta}{\partial \rho} = -\frac{\eta c}{\xi^2}\frac{\partial \xi}{\partial \rho}, \quad \frac{\partial \xi}{\partial \rho} = -\frac{1+\eta}{\rho^2}. \tag{4.64}$$

It can be shown that

$$a^2 = \frac{\left(\frac{\partial f}{\partial \rho}\right)}{\left(\frac{\partial f}{\partial P_L}\right)} = \left(\frac{\partial P_{\mathrm{L}}}{\partial \rho}\right) = \frac{3\beta^2\gamma - \delta\epsilon}{3\beta^2\theta - \delta\lambda}. \tag{4.65}$$

Additionally,

$$\gamma = \frac{(1+\eta)\eta c}{\rho^2\xi^2} \tag{4.66}$$

$$\epsilon = \frac{1+\eta}{\rho^2\xi^2} \tag{4.67}$$

$$\theta - 1 = \frac{\eta c * \exp\left[\frac{P_{\mathrm{A}} - P_{\mathrm{L}}}{K}\right]}{K\rho_{\mathrm{LA}}\xi^2}. \tag{4.68}$$

And

$$\lambda = \frac{\exp\left[\frac{P_{\mathrm{A}} - P_{\mathrm{L}}}{K}\right]}{K\rho_{\mathrm{LA}}\xi^2}. \tag{4.69}$$

Referring to Figs. 4.5 and 4.6, it is apparent that the speed of sound is determined solely by bubble radius for $r < 10^{-7}$ in. and solely by liquid pressure when $r > 10^{-4}$ in. [6]. In the intermediate range, the speed of sound is dependent both on bubble radius and liquid pressure.

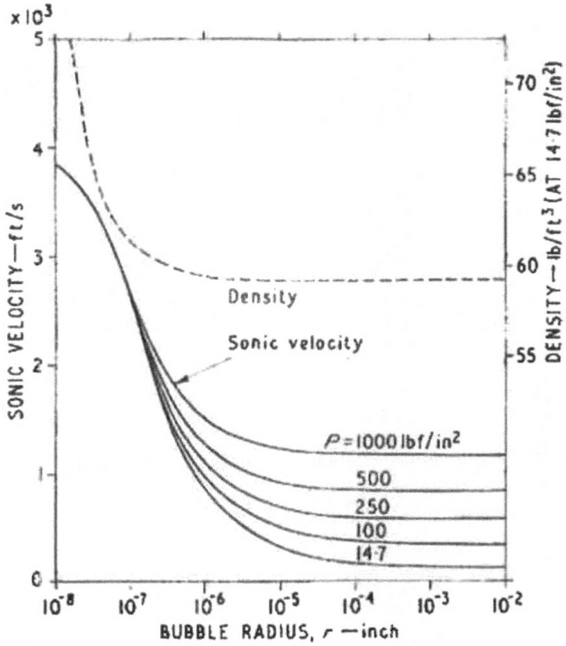

Fig. 4.5 Sonic velocity versus bubble radius (void fraction, 0.05) [6]

4.6 Speed of Sound for Gas Mixtures

We saw in Chap. 1 that when pressure increases for a fixed temperature or temperature decreases for a fixed pressure, then some of the assumptions that ideal gas behavior is based on are no longer valid. To address these concerns, cubic equations of state (EOS) were developed that mimic the behavior of pressure versus v for real gases; further, these EOS's address other concerns.

The cubic EOS can be extended to gas mixtures through what are called mixing rules; the mixing rule utilized in this section is known as van der Waals one-fluid mixing rules [10]. Here's how it works.

The basic idea is that we get a weight average for the cubic EOS parameters $\{a, b\}$ based on the molar concentration of each of two substances; please note these ideas can be extended to more than a binary mixture but won't be explored here.

The mixture coefficients $\{a_{\text{mix}}, b_{\text{mix}}\}$ are defined [10] as

$$a_{\text{mix}} = \begin{bmatrix} \chi_1 & \dots & \chi_n \end{bmatrix} \begin{bmatrix} a_{11} & \cdots & a_{1n} \\ \vdots & \ddots & \vdots \\ a_{n1} & \cdots & a_{nn} \end{bmatrix} \begin{bmatrix} \chi_1 \\ \dots \\ \chi_n \end{bmatrix}. \tag{4.70}$$

And

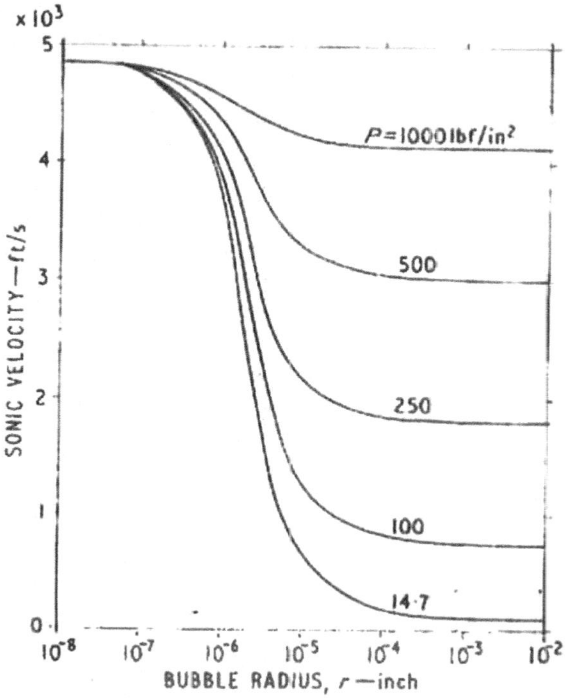

Fig. 4.6 Sonic velocity versus bubble radius (mass fraction, 1e−4) [6]

$$b_{\text{mix}} = \begin{bmatrix} \chi_1 & \dots & \chi_n \end{bmatrix} \begin{bmatrix} b_1 \\ \dots \\ b_n \end{bmatrix}, \tag{4.71}$$

where

$$a_{ij} = a_{ji} = \sqrt{a_i a_j}. \tag{4.72}$$

And x_i is the mole fraction for pure component i, $\{a_i, b_i\}$ are the coefficients for pure component i, and n is the number of components.

An example of this method for a binary mixture is given below.

Example 4.1 Speed of sound for mixtures

In this example, the effects of molar concentration of a binary mixture of methane and air at 1200 K and with a specific volume of $0.1\ \frac{l}{\text{mole}}$ are explored using three models for equation of state:

1. Van der Waal.
2. Redlich–Kwong.
3. Ideal gas.

And a cubic model for specific heat will be utilized. As can be seen in Fig. 4.7, the effects of molar concentration on the coefficients $\{a, b\}$ associated with the equations of state are determined. Figure 4.8 is a table where the effects of molar concentration on pressure are determined and the results are graphically illustrated in Fig. 4.9.

In Fig. 4.10, cubic models for specific heat are determined for 1200 K and the ratio $\frac{R_g}{c_v}$ is determined for various molar concentrations. This model $\left(1+\frac{R_g}{c_v}\right)$ becomes the basis to γ assuming ideal gas behavior as was discussed in Chap. 3.

Substance	a(VW)	b(VW)	a(RK)	b(RK)	Alpha(T)
Methane	2.24	0.04	31.39	0.03	0.03
Air	1.33	0.04	15.45	0.03	0.03

T	[K]	1200
nu(mix)	[L/mole]	0.1
R(u)	Atm/mole	0.08206

VW EOS

y(CH4) =	0.7		
y(Air) =	0.3		
y =	0.7	0.3	
A =	2.24	1.72	
	1.72	1.33	
y' =	0.7		
	0.3		
a(mix) =	2.09		1.94
	1.60		
B =	0.04		
	0.04		
b(mix) =	0.041		
P(VW)	[atm]	1463.9	

RK EOS

y(CH4) =	0.7		
y(Air) =	0.3		
y =	0.7	0.3	
A =	31.39	22.02	
	22.02	15.45	
y' =	0.7		
	0.3		
a(mix) =	28.58		26.02
	20.05		
B =	0.03		
	0.03		
b(mix)	0.028	alpha =	0.029
P(RK) =	[atm]	1311.9	

$$a_{mix} = [\chi_1 \cdots \chi_n]\begin{bmatrix} a_{11} & \cdots & a_{1n}\\ \vdots & \ddots & \vdots \\ a_{n1} & \cdots & a_{nn}\end{bmatrix}\begin{bmatrix}\chi_1\\ \cdots \\ \chi_n\end{bmatrix}$$

$$a_{ij} = a_{ji} = \sqrt{a_i a_j}$$

$$b_{mix} = [\chi_1 \cdots \chi_n]\begin{bmatrix} b_1\\ \cdots \\ b_n\end{bmatrix}$$

$$P = P_R + P_A = \frac{R_u T}{v-b} - \frac{a}{v^2}$$

$$P = \frac{R_u T}{v-b} - \frac{\alpha a}{v(v+b)}$$

Fig. 4.7 Mixing laws

Y(CH4)	Y(Air)	a(vw)	b(vw)	a(RK)	b(RK)	alpha	P(vw) [atm]	P(RK) [atm]
0.1	0.9	1.41	0.037	16.79	0.026	0.029	1418	1284
0.2	0.8	1.49	0.037	18.19	0.026	0.029	1426	1288
0.3	0.7	1.58	0.038	19.64	0.027	0.029	1433	1293
0.4	0.6	1.66	0.039	21.15	0.027	0.029	1441	1298
0.5	0.5	1.75	0.039	22.72	0.027	0.029	1448	1303
0.6	0.4	1.85	0.04	24.34	0.028	0.029	1456	1307
0.7	0.3	1.94	0.041	26.02	0.028	0.029	1464	1312
0.8	0.2	2.04	0.041	27.75	0.029	0.029	1472	1317
0.9	0.1	2.14	0.042	29.55	0.029	0.029	1480	1321
1	0	2.24	0.043	31.39	0.029	0.029	1488	1326

Fig. 4.8 Calculation of pressure for van der Waal and Redlich–Kwong gases

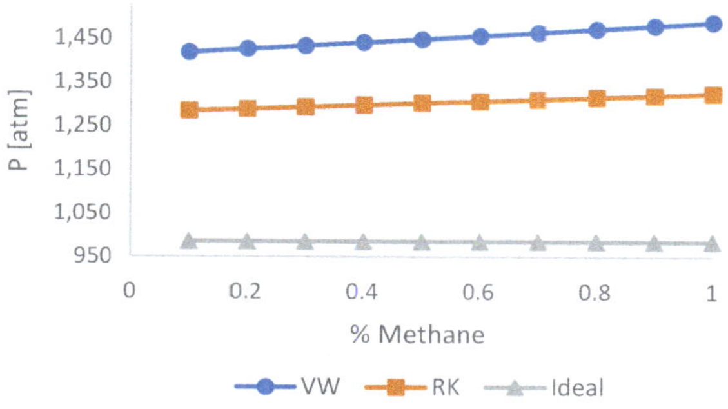

Fig. 4.9 Pressure versus molar concentration

	γ	β_0	β_1	β_2	β_3
Methane	1.299	1.2	3.25	0.75	-0.71
Air	1.4	1.05	-0.365	0.85	-0.39

T	[K]	1200
nu(mix)	[L/mole]	0.1
R(u)	[L-Atm/mole-K]	0.08206
Omega	[K]	1.2

$$C_p = \beta_0 + \beta_1\theta + \beta_2\theta^2 + \beta_3\theta^3$$

$$\theta = \frac{T[K]}{1000}$$

Y(CH4)	Y(Air)	C(v,Methane) [kJ/kg-K]	C(v, Air) [kJ/kg-K]	C(v,mix) [kJ/kg-K]	MW(Methane) [gm/mole]	MW(Air) [gm/mole]	MW(mix) [gm/mole]	R(U) [L-Atm/mole-K]	R(g) [kJ/kg-K]	R(g)/C(v)
0.1	0.9	3.81	0.83	1.1	16.01	29	27.7	0.08206	0.30	0.266
0.2	0.8	3.81	0.83	1.4	16.01	29	26.4	0.08206	0.31	0.221
0.3	0.7	3.81	0.83	1.7	16.01	29	25.1	0.08206	0.33	0.192
0.4	0.6	3.81	0.83	2.0	16.01	29	23.8	0.08206	0.35	0.173
0.5	0.5	3.81	0.83	2.3	16.01	29	22.5	0.08206	0.37	0.159
0.6	0.4	3.81	0.83	2.6	16.01	29	21.2	0.08206	0.39	0.150
0.7	0.3	3.81	0.83	2.9	16.01	29	19.9	0.08206	0.42	0.143
0.8	0.2	3.81	0.83	3.2	16.01	29	18.6	0.08206	0.45	0.139
0.9	0.1	3.81	0.83	3.5	16.01	29	17.3	0.08206	0.48	0.137
1	0	3.81	0.83	3.8	16.01	29	16.0	0.08206	0.52	0.136

Fig. 4.10 Determination of gamma as a function of molar concentration

In Fig. 4.11, the speed of sound for both a van der Waal and a Redlich–Kwong gas is determined to include a model for γ. The results of this analysis are given in Fig. 4.12, and for comparison, an ideal gas speed of sound for differing molar concentrations is also included. As can be seen in Fig. ZZ, not properly addressing the non-ideal behavior can result in a speed of sound calculation well below actual conditions. For a large, reduced temperature ($T_r > 6.3$) and a large, reduced pressure ($20 < P_r < 30$), it is obvious that the Z value will be far from unity.

4.7 Omega Method

When the system of concern is homogenous, then the following method known as the omega method [12] is appropriate. The omega method is additionally the subject of Chap. 9 where the basic ideal given here will be extended to either flashing, non-flashing pipes and flashing, non-flashing nozzles and show how this method can be extended to multi-component systems and non-ideal cases.

Y(CH4)	Y(Air)	a(vw)	b(vw)	R(g)/C(v)	[1+R/C(V)]	R(g) [J/kg-K]	(nu^2*RT)/(nu-b)^2	c(vw)^2	C(vw) [m/s]
0.1	0.9	1.41	0.037	0.266	1.266	300.1	907259.9	1148502.32	1071.7
0.2	0.8	1.49	0.037	0.221	1.221	314.8	951897.8	1161930.88	1077.9
0.3	0.7	1.58	0.038	0.192	1.192	331.1	1033711.3	1232111.58	1110.0
0.4	0.6	1.66	0.039	0.173	1.173	349.2	1126156.3	1320486.49	1149.1
0.5	0.5	1.75	0.039	0.159	1.159	369.4	1191158.6	1380631.94	1175.0
0.6	0.4	1.85	0.04	0.150	1.150	392.0	1306613.1	1502067.88	1225.6
0.7	0.3	1.94	0.041	0.143	1.143	417.6	1439456.1	1645386.63	1282.7
0.8	0.2	2.04	0.041	0.139	1.139	446.7	1539942.6	1753769.94	1324.3
0.9	0.1	2.14	0.042	0.137	1.137	480.2	1713090.5	1947110.63	1395.4
1	0	2.24	0.043	0.136	1.136	519.2	1917640.6	2178705.56	1476.0

Y(CH4)	Y(Air)	a(RK)	b(RK)	alpha	R(g)/C(v)	R(g) [J/kg-K]	[1+R/C(V)]	(nu^2*RT)/(nu-b)^2	c(RK)^2	c(RK) [m/s]
0.1	0.9	16.79	0.026	0.029	0.266	300.1	1.266	657,581	832,451	912.4
0.2	0.8	18.19	0.026	0.029	0.221	314.8	1.221	689,935	842,184	917.7
0.3	0.7	19.64	0.027	0.029	0.192	331.1	1.192	745,653	888,784	942.8
0.4	0.6	21.15	0.027	0.029	0.173	349.2	1.173	786,344	922,053	960.2
0.5	0.5	22.72	0.027	0.029	0.159	369.4	1.159	831,732	964,050	981.9
0.6	0.4	24.34	0.028	0.029	0.150	392.0	1.150	907,370	1,043,121	1021.3
0.7	0.3	26.02	0.028	0.029	0.143	417.6	1.143	966,579	1,104,877	1051.1
0.8	0.2	27.75	0.029	0.029	0.139	446.7	1.139	1,063,388	1,211,064	1100.5
0.9	0.1	29.55	0.029	0.029	0.137	480.2	1.137	1,143,193	1,299,380	1139.9
1	0	31.39	0.029	0.029	0.136	519.2	1.136	1,235,948	1,404,227	1185.0

Fig. 4.11 Determination of speed of sound for van der Waal and Redlich–Kwong gases

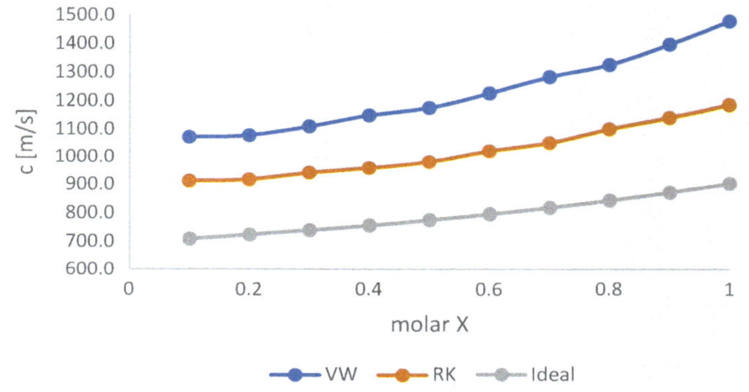

Fig. 4.12 Speed of sound versus molar concentration

By definition,

$$c^2 = v^2\left[-\left(\frac{\partial v}{\partial P}\right)_{\mathrm{s}}^{-1}\right], \tag{4.73}$$

where the following thermodynamic relationships will be utilized to develop a homogeneous, two phase sonic velocity (Clausius–Clapeyron relationships)

$$\frac{\mathrm{d}P}{\mathrm{d}T} = \frac{s_{\mathrm{fg}}}{v_{\mathrm{fg}}},\ s_{\mathrm{fg}} = \frac{h_{\mathrm{fg}}}{T},\ \frac{\mathrm{d}s_{\mathrm{f}}}{\mathrm{d}T} \approx \frac{c_{\mathrm{pf}}}{T}. \tag{4.74}$$

Starting from the definition for specific volume of a mixture of gas and liquid

$$\nu_{\text{mix}} = \nu_{\text{l}} + x(\nu_{\text{g}} - \nu_{\text{l}}). \tag{4.75}$$

And taking the derivation of Eq. 4.75 with respect to P results in

$$\begin{aligned}\left(\frac{\partial \nu_{\text{mix}}}{\partial P}\right)_{\text{s}} &= \frac{\partial \nu_{\text{l}}}{\partial P} + x\frac{\partial}{\partial P}[\nu_{\text{g}} - \nu_{\text{l}}] + [\nu_{\text{g}} - \nu_{\text{l}}]\frac{\partial x}{\partial P} \\ &= x\frac{\partial \nu_{\text{g}}}{\partial P} + \nu_{\text{fg}}\frac{\partial x}{\partial P} + (1-x)\frac{\partial \nu_{\text{l}}}{\partial P}\end{aligned}, \tag{4.76}$$

where we'll assume the liquid phase is an incompressible fluid and so $\frac{\partial \nu_{\text{l}}}{\partial P} \approx 0$.

Given below are derivation for $\frac{\partial \nu_{\text{g}}}{\partial P}$ and $\frac{\partial x}{\partial P}$.

Assuming ν_{g} acts ideal

$$\nu_{\text{g}} = \frac{RT}{P}. \tag{4.77}$$

And taking the derivative of both sides with respect to P results in

$$\frac{\text{d}\nu_{\text{g}}}{\text{d}P} = R\frac{\text{d}}{\text{d}P}\left(\frac{T}{P}\right) = R\left[\frac{1}{P}\frac{\text{d}T}{\text{d}P} - \frac{1}{P^2}T\right]. \tag{4.78}$$

And

$$\frac{R}{P} = \frac{\nu_{\text{g}}}{T} \text{ and } -\frac{1}{P}\left(\frac{RT}{P}\right) = -\frac{\nu_{\text{g}}}{P}. \tag{4.79}$$

And

$$\frac{\text{d}P}{\text{d}T} = \frac{s_{\text{fg}}}{\nu_{\text{fg}}} = \frac{h_{\text{fg}}}{T\nu_{\text{fg}}}. \tag{4.80}$$

Substituting Eqs. 4.79 and 4.80 into 4.78 results in

$$\frac{\text{d}\nu_{\text{g}}}{\text{d}P} = -\frac{\nu_{\text{g}}}{P} + \frac{\nu_{\text{g}}\nu_{\text{fg}}}{h_{\text{fg}}}. \tag{4.81}$$

And

$$\left(\frac{\text{d}x}{\text{d}P}\right)_{\text{s}} = \frac{\left(\frac{\text{d}x}{\text{d}T}\right)_{\text{s}}}{\left(\frac{\text{d}P}{\text{d}T}\right)}. \tag{4.82}$$

And the Clausius–Clapeyron relationships is

$$\frac{\text{d}P}{\text{d}T} = \frac{h_{\text{fg}}}{T\nu_{\text{fg}}}. \tag{4.83}$$

And

$$s_{\text{mix}} = s_{\text{f}} + x s_{\text{fg}}. \tag{4.84}$$

And taking the derivative of both sides with respect to T for an isentropic process results in

$$0 = \frac{\mathrm{d}s_{\text{mix}}}{\mathrm{d}T} = \frac{\mathrm{d}s_{\text{f}}}{\mathrm{d}T} + x\frac{\mathrm{d}s_{\text{fg}}}{\mathrm{d}T} + s_{\text{fg}}\frac{\mathrm{d}x}{\mathrm{d}T} \tag{4.85}$$

or

$$s_{\text{fg}}\frac{\mathrm{d}x}{\mathrm{d}T} = \frac{h_{\text{fg}}}{T}\frac{\mathrm{d}x}{\mathrm{d}T} = -\frac{\mathrm{d}s_{\text{f}}}{\mathrm{d}T} - x\frac{\mathrm{d}s_{\text{fg}}}{\mathrm{d}T} = -\frac{c_{\text{pf}}}{T} - x\frac{\mathrm{d}}{\mathrm{d}T}\left[\frac{h_{\text{fg}}}{T}\right], \tag{4.86}$$

where h_{fg} is constant with respect to T and so

$$\frac{\mathrm{d}x}{\mathrm{d}T} = -\left\{\frac{c_{\text{pf}}}{T} + x\frac{\mathrm{d}}{\mathrm{d}T}\left[\frac{h_{\text{fg}}}{T^2}\right]\right\}\frac{T}{h_{\text{fg}}}. \tag{4.87}$$

And

$$\frac{\mathrm{d}x}{\mathrm{d}T} = -\frac{c_{\text{pf}}}{h_{\text{fg}}} + \frac{x}{T}. \tag{4.88}$$

And

$$\begin{aligned}\left(\frac{\mathrm{d}x}{\mathrm{d}P}\right)_{\text{s}} &= \frac{\left(\frac{\mathrm{d}x}{\mathrm{d}T}\right)_{\text{s}}}{\left(\frac{\mathrm{d}P}{\mathrm{d}T}\right)} = \frac{-\frac{c_{\text{pf}}}{h_{\text{fg}}} + \frac{x}{T}}{\frac{h_{\text{fg}}}{T v_{\text{fg}}}} \\ &= -\frac{c_{\text{pf}} v_{\text{fg}} T}{h_{\text{fg}}^2} + \frac{x v_{\text{fg}}}{h_{\text{fg}}} = \frac{v_{\text{fg}}}{h_{\text{fg}}}\left[x - \frac{c_{\text{pf}} T}{h_{\text{fg}}}\right]\end{aligned}. \tag{4.89}$$

Substitution of Eqs. 4.81 and 4.89 into Eq. 4.76 results in

$$\left(\frac{\partial v_{\text{mix}}}{\partial P}\right)_{\text{s}} = x\frac{\partial v_{\text{g}}}{\partial P} + v_{\text{fg}}\frac{\partial x}{\partial P} = -\frac{v_{\text{g}} x}{P} + \frac{v_{\text{g}} v_{\text{fg}} x}{h_{\text{fg}}} + v_{\text{fg}}\left\{\frac{v_{\text{fg}}}{h_{\text{fg}}}\left[x - \frac{c_{\text{pf}} T}{h_{\text{fg}}}\right]\right\} \tag{4.90}$$

or

$$\left(\frac{\partial v}{\partial P}\right)_{\text{s}} = \frac{-x v_{\text{g}}}{P}\left[1 - \frac{P v_{\text{fg}}}{h_{\text{fg}}}\left(1 + \frac{v_{\text{fg}}}{v_{\text{g}}}\right)\right] - c_{\text{pf}} T\left(\frac{v_{\text{fg}}}{h_{\text{fg}}}\right)^2. \tag{4.91}$$

And a dimensionless version of the above equation along with the following observation, $v_{fg} \approx v_g$ in terms of c is

$$\frac{c}{\sqrt{Pv}} = \left\{ \frac{xv_g}{v} \left[1 - 2\frac{Pv_{fg}}{h_{fg}} \right] + \frac{c_{pf}TP}{v} \left(\frac{v_{fg}}{h_{fg}} \right)^2 \right\}^{-1/2}. \quad (4.92)$$

And setting ω equal to

$$\omega = \alpha \left[1 - 2\frac{Pv_{fg}}{h_{fg}} \right] + \frac{c_{pf}TP}{v} \left(\frac{v_{fg}}{h_{fg}} \right)^2 \quad (4.93)$$

results in

$$c^* = \frac{c}{\sqrt{Pv}} = \frac{1}{\sqrt{\omega}}. \quad (4.94)$$

Example 4.2 Speed of sounds using omega method

Using Leung's method as given by Eqs. 4.92, 4.93 and 4.94, determine the speed of sound of saturated water at 100 C and quality equal to 50%. Given below is the solution and we can see that the speed of sound for this saturated water mixture is 298 m/s (Fig. 4.13).

Fig. 4.13 Solution for Example 4.2

Subtance		Water
T(0)	[C]	100
T(0)	[K]	373
P(0)	[kPa]	101.3
x(0)		50%
Gamma		1.327
v(l,0)	[m^3/kg]	0.001044
v(g,0)	[m^3/kg]	1.6729
v(fg0)	[m^3/kg]	1.671856
v(0)	[m^3/kg]	0.836972
c(f0)	[kJ/kg-K]	4.18
h(fg0)	[kJ/kg]	2257.03
Alpha		0.999

Omega(1)		0.85
Omega(2)		0.10
Omega		0.95
c/sqrt(P*nu)		1.02
c	[m/s]	298

$$\omega = \alpha\left[1 - 2\frac{Pv_{fg}}{h_{fg}}\right] + \frac{c_{pf}TP}{v}\left(\frac{v_{fg}}{h_{fg}}\right)^2$$

$$c^* = \frac{c}{\sqrt{Pv}} = \frac{1}{\sqrt{\omega}}$$

References

1. F.T. Barclay, T.J. Ledwidge, G.C. Cornfield, Some experiments on sonic velocity in two-phase one-component mixtures and some thoughts on the nature of two-phase critical flow. Proc. Inst. Mech. Eng. **183**(Part 3C), 102–107 (1969)
2. E.L. Carstensen, L.L. Foldy, Propagation of sound through a liquid containing bubbles. J. Acoust. Soc. Am. **19**(3), 481–501 (1947)

3. A.L. Davies, The speed of sound in mixtures of water and steam (1965)
4. H.B. Karplus, The velocity of sound in a liquid containing gas bubbles, in *Armour Research Foundation Project No. A-097*, 11 June 1958
5. S.W. Kieffer, Sound speed in liquid-gas mixtures: water-air and water steam. J. Geophys. Res. **82**(20), 2895–2904 (1977)
6. D. McWilliams, R.K. Duggins, Speed of sound in bubbly liquids. Proc. Inst. Mech. Eng. **183**(Part 3C), 102–107 (1969)
7. D.J. Picard, P.R. Bishnoi, Calculation of the thermodynamic sound velocity in two-phase multicomponent fluids. Int. J. Multiphase Flow **13**(3), 295–308 (1987)
8. A.B. Wood, *A Textbook of Sound* (The MacMillan Company, New York, NY, 1930)
9. https://en.wikipedia.org/wiki/Multiphase_flow. Accessed 7 May 2024
10. J.M. Smith, H.C. Van Ness, M.M. Abbott, M.T. Swihart, *Introduction to Chemical Engineering Thermodynamics* (McGraw Hill, New York, NY, 2005)
11. K. Li, Q. Wang, M. Wang, Three-dimensional visualization of gas-water two-phase flow based on bubble mapping method and size projection algorithm. Flow Meas. Instrum. **69**(3) (2019)
12. J.C. Leung, On the application of the method of Landau and Lifshitz to sonic velocities in homogenous two-phase mixtures. Trans. ASME **118**, 186–188 (1996)

Chapter 5
Speed of Sound in Liquids

Abstract This chapter begins by showing how the Tait equation of state to describe states associated with a liquid comes from a linear relationship between bulk modulus and pressure for an isothermal data set. The chapter then goes on to show several variations of the original Tait equation to include models for equations of state in salt water and liquid mixtures based on corresponding states. From these equations of states, it is shown how speed of sound for a pure liquid and liquid mixture can be derived. The chapter ends with a discussion on how one challenge associated with determining speed of sound through a liquid is how to determine a model for γ, which has been shown to related the two definitions of speed of sound—adiabatic and isothermal. This discussion includes a comparison of a model for speed of sound and empirical data provided from the National Institute of Standards and Testing where the discrepancy between the model and empirical data is exactly the absence of γ.

Keywords Tait equation of state · Isothermal and isentropic compressibility · Bulk modulus · Speed of sound through liquids and speed of sound through liquid mixtures

5.1 Preview

In this chapter, we'll develop equations for the speed of sound in liquids to include speed of sound in water at various temperatures, pressures, and salinity concentrations; speed of sound through liquid mixtures; and speed of sound in liquid hydrocarbons.

In Sect. 5.2, an equation will be developed that relates densities to pressures for liquids and some solids. This relationship is developed out of the work of Tait [1–4]. In Sect. 5.3, the basic Tait equation will be extended to slightly polar hydrocarbons and liquid mixtures through the idea of corresponding states; the concept of corresponding states is the basis to Lee–Kessler compressibility charts [5].

H. Foust, *Speed of Sound and Critical Flow*, Shock Wave and High Pressure Phenomena, https://doi.org/10.1007/978-981-97-5918-7_5

In Sect. 5.4, the speed of sound will be presented based on the relationships derived in Sects. 5.2 and 5.3 for water, seawater, and slightly polar liquid hydrocarbons.

5.2 Density in Liquids

The volume expansion of a substance can be expressed as

$$\mathrm{d}\nu = \left(\frac{\partial \nu}{\partial T}\right)_{\mathrm{P}} \mathrm{d}T + \left(\frac{\partial \nu}{\partial P}\right)_{\mathrm{T}} \mathrm{d}P, \tag{5.1}$$

where the volume expansivity (β_{p}) is defined as

$$\beta_{\mathrm{p}} = \frac{1}{\nu}\left(\frac{\partial \nu}{\partial T}\right)_{\mathrm{P}}. \tag{5.2}$$

And the isothermal compressibility (β_{T}) is

$$\beta_{\mathrm{T}} = \frac{-1}{\nu}\left(\frac{\partial \nu}{\partial P}\right)_{\mathrm{T}}. \tag{5.3}$$

Along an isentropic path (β_{s})

$$\beta_{\mathrm{s}} = \frac{-1}{\nu}\left(\frac{\partial \nu}{\partial P}\right)_{\mathrm{s}}. \tag{5.4}$$

And the relative change in ν can be defined as

$$\frac{\mathrm{d}\nu}{\nu} = \beta_{\mathrm{P}}\mathrm{d}T - \beta_{\mathrm{T}}\mathrm{d}P. \tag{5.5}$$

So that for a *constant pressure* process

$$\frac{\mathrm{d}\nu}{\nu} = \beta_{\mathrm{p}}\mathrm{d}T \tag{5.6}$$

or

$$\ln\left(\frac{\nu}{\nu_0}\right) = \int_{T_0}^{T} \beta_{\mathrm{p}}\mathrm{d}T. \tag{5.7}$$

Equation 5.7 shows us that increases in temperature increases $\frac{\mathrm{d}\nu}{\nu}$ along a constant pressure line.

A *constant temperature* process is

$$\frac{\mathrm{d}\nu}{\nu} = -\beta_{\mathrm{T}}\mathrm{d}P \tag{5.8}$$

or

$$\ln\left(\frac{\nu_0}{\nu}\right) = \int_{P_0}^{P} \beta_{\mathrm{T}}\mathrm{d}P. \tag{5.9}$$

Equation 5.9 shows us that increases in pressure decreases $\frac{\mathrm{d}\nu}{\nu}$ along a constant temperature line.

We define bulk modulus as

$$K_{\mathrm{T}} = -\nu\frac{\partial P}{\partial \nu}_{\mathrm{T}} \tag{5.10}$$

or

$$K_{\mathrm{s}} = -\nu\frac{\partial P}{\partial \nu}_{\mathrm{S}} \tag{5.11}$$

where bulk modulus is usually defined along an isothermal path [2].

Please note that bulk modulus and isothermal compressibility are inverses.

This section will develop equations of state for liquids based on an empirical relationship between the isothermal compressibility and experimental results [2] of the form

$$\ln\left(\frac{\nu_0}{\nu}\right) = \int_{P_0}^{P} \beta_{\mathrm{T}}\mathrm{d}P \approx \int_{P_0}^{P} \frac{1}{K_T}\mathrm{d}P \approx \int_{P_0}^{P} \frac{\mathrm{d}P}{K_0 + nP}. \tag{5.12}$$

Please note K_0 is the bulk modulus at standard condition and n is a constant associated with the substance

The left-hand side of Eq. 5.12 comes from Eq. 5.9, and the right-hand side of Eq. 5.12 comes from the empirical relationship given in Example 5.1.

Example 5.1 Experimental data for bulk modulus

Determine the relationship between pressure and bulk modulus for DEHP, which is a liquid added to plastics to increase flexibility, where data is given for two temperatures [6].

Figure 5.1 provides a clear relationship between K versus P, which is given as

$$K_{\mathrm{T}} = -\nu\left(\frac{\partial P}{\partial \nu}\right)_{\mathrm{T}} = K_0 + nP, \tag{5.13}$$

where K_0 is the bulk modulus at standard conditions and n is a determined coefficient.

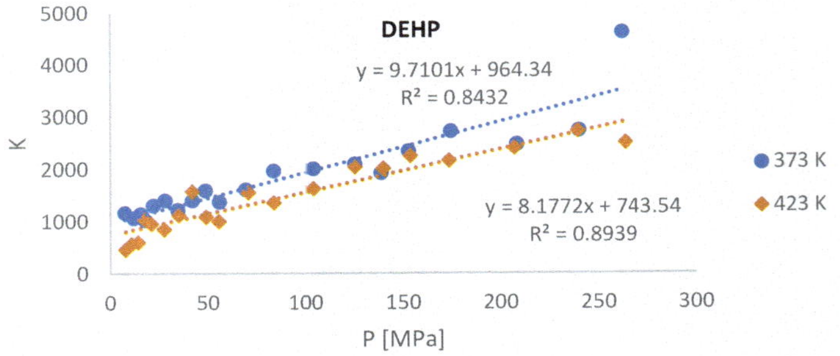

Fig. 5.1 Bulk modulus versus pressure [6]

It is noted from Fig. 5.1 that temperature has an effect on the bulk modulus (K_T), which is not as significant as the effects of pressure.

Integrating Eq. 5.12 results in

$$\left(\frac{\nu_0}{\nu}\right)^n = \frac{K_0 + nP}{K_0 + nP_0}. \tag{5.14}$$

And

$$\frac{K_0}{n}\left[\left(\frac{\nu_0}{\nu}\right)^n - 1\right] + P_0\left(\frac{\nu_0}{\nu}\right)^n = P \tag{5.15}$$

or

$$\frac{K_0}{n}\left[\left(\frac{\nu_0}{\nu}\right)^n - 1\right] + P_0 \approx P \tag{5.16}$$

5.2.1 *Tait's Equation of State*

The original form of the Tait equation, which was published in 1888 [2, 3, 5] and is given as

$$\frac{-1}{\nu_0}\frac{\Delta\nu}{\Delta P} = \frac{A}{\pi + (P - P_0)}, \tag{5.17}$$

where $\{\upsilon_0, P_0\}$ are the initial specific volume and pressure, $\{A, \pi\}$ are fitted parameters, and P is the final pressure.

The original Tait equation was replaced by the following form [2, 3, 5]

$$\frac{-1}{\nu}\frac{\mathrm{d}\nu}{\mathrm{d}P} = \frac{A}{\nu(B + P)} \tag{5.18}$$

or in terms of isothermal compressibility

$$\beta_T = \frac{-1}{\nu}\left(\frac{d\nu}{dP}\right)_T = \frac{0.4343C}{\nu(B+P)}. \tag{5.19}$$

And integrated becomes

$$\frac{\nu}{\nu_0} = 1 - \frac{C}{\nu_0}\log_{10}\left(\frac{B+P}{B+P_0}\right) \tag{5.20}$$

5.2.2 *Murnaghan–Tait Equation of State*

Another form of the *Tait* equation is the *Murnaghan–Tait* equation [3] given in terms of specific volume as

$$\frac{\nu}{\nu_0} = \left[1 + \frac{n}{K_0}(P - P_0)\right]^{-1/n}, \tag{5.21}$$

where $\{\upsilon_0, P_0\}$ are the initial specific volume and pressure, $\{n, K_0\}$ are fitted parameters, and P is the final pressure

In a pressure form, this equation is

$$P \approx \frac{K_0}{n}\left[\left(\frac{\nu_0}{\nu}\right)^n - 1\right] + P_0 \tag{5.22}$$

which was given earlier as Eq. 5.16

Or

$$\frac{P}{P_0} \approx (B+1)\left(\frac{\rho}{\rho_0}\right)^n - B \tag{5.23}$$

where

$$B = \frac{K_0}{nP_0}. \tag{5.24}$$

And n is a constant associated with the material.

Example 5.2 Determination of $\{n, B\}$ for experimental data for Mercury

Determine the coefficients $\{n, B\}$ associated with the *Murnaghan–Tait* equation applied to the following data provided for Mercury at 20 °C [4]—see Table 5.1.

Utilizing a nonlinear optimization tool (Solver) within Excel the coefficient are determined to be $\{n; B\} = \{6; 41,000\}$ and the comparison of the experimental

Table 5.1 Density data for Mercury [4]

P (atm)	Rho (kg/m^3)
1	13,545
500	13,573
1000	13,600
1500	13,625
2000	13,653

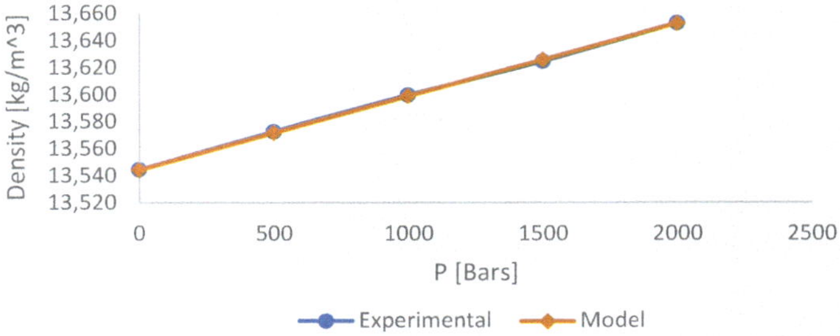

Fig. 5.2 Density versus pressure for Mercury at 20 °C

results and model are given in Fig. 5.2 where the model has a high fit to the provided experimental data.

5.2.3 Density of Various Liquids Under Pressure

The National Institute of Standards and Testing provides a wealth of thermodynamic properties for liquids and gases through the NIST Webbook [7]. Utilizing isothermal data for a particular liquid provides the basis of models for density through the use of the MT Equation of State, as we'll see in the next section, the speed of sound can be determined by either

1. Derivative from MT Equation of State for $\left(\frac{dP}{d\rho}\right)_T$ and a model for γ.
2. Determination of isentropic bulk modulus and density.

5.2.4 Density of Salt Water Under Various Conditions

Yet another form of the *Tait* equation that allows for changes in salinity is called the *Tumlirz–Tammann–Tait* equation of state [3].

$$\nu = \nu_\infty - k_1 S + \frac{\lambda}{P_0 + k_2 S + P}, \tag{5.25}$$

where ν_∞ when P_∞, $\{k_1, k_2 P_0\}$ are fitted parameters, S is the salinity, and $\{\nu, P\}$ are the current specific volume [cm^3/gm] and pressure [bars].

When $S = 0$, a simpler form of Eq. 5.25 results and is given as

$$\nu = \nu_\infty + \frac{\lambda}{P_0 - P}, \tag{5.26}$$

where the parameters are temperature dependent and have the following values

$$\lambda = a_1 + a_2 T + a_3 T^2 + a_4 T^4 + a_5 T^5 \tag{5.27}$$

$$P_0 = b_1 + b_2 T + b_3 T^2 + b_4 T^4 + b_5 T^5 \tag{5.28}$$

And

$$\nu_\infty = c_1 + c_2 T + c_3 T^2 + c_4 T^3 + c_5 T^4 + c_6 T^5 + c_7 T^6 + c_8 T^7 + c_9 T^8 + c_{10} T^9, \tag{5.29}$$

where the values of the parameters are given in Table 5.2 [3].

Example 5.3 Effects of pressure and temperature on freshwater

Utilizing the freshwater form of the Tumlirz–Tammann–Tait equation (Eq. 5.26) to show the effects of pressure and temperature on density.

The results of an analysis for temperatures 20, 40 and 60 °C for a range of pressure from 1000 to 10,000 Bars are given as Fig. 5.3. White [4] has noted that the change in density between the surfaces of our oceans to the deepest part of the oceans at crushing pressures (≈ 1100 atmospheres) is only a few percentages of difference.

The effects of salinity on density [8] can be seen in Fig. 5.4.

Table 5.2 Parameters associated with Eqs. 5.27, 5.28, and 5.29 [8]

a(1)	1788.316	**b(1)**	5918.499	**c(1)**	0.698055
a(2)	21.55053	**b(2)**	58.05267	**c(2)**	– 7.44E – 04
a(3)	– 0.4695911	**b(3)**	– 1.12533	**c(3)**	3.70E – 05
a(4)	3.10E – 03	**b(4)**	6.61E – 03	**c(4)**	– 6.32E – 07
a(5)	– 7.38E – 06	**b(5)**	– 1.47E – 05	**c(5)**	9.83E – 09
				c(6)	– 1.20E – 10
				c(7)	1.01E – 12
				c(8)	– 5.44E – 15
				c(9)	1.80E – 17
				c(10)	– 2.30E – 20

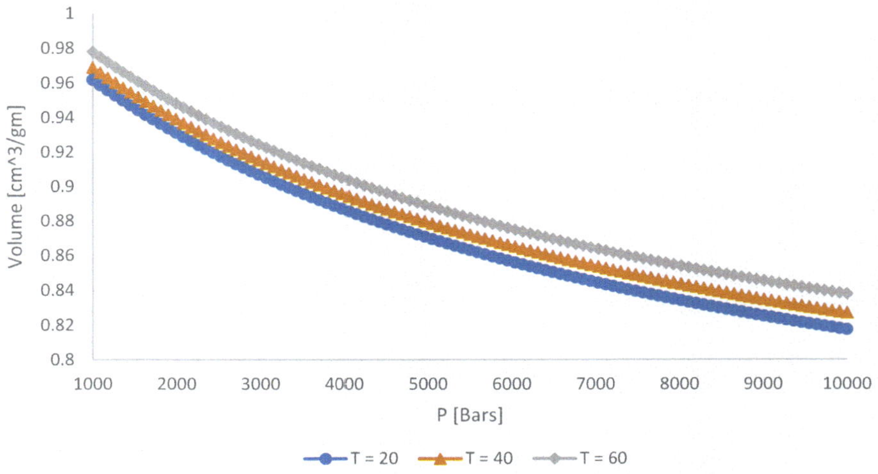

Fig. 5.3 Specific volume versus pressure for three temperatures

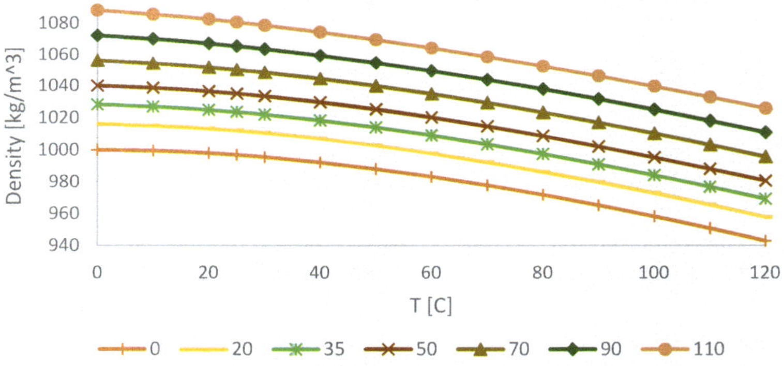

Fig. 5.4 Density versus temperature for various salinities [gm/kg]

The form of equation to be utilized for determining speed of sound through saline water is

$$K = -\nu\frac{\partial P}{\partial \nu} = \frac{\nu\lambda}{(\nu - \nu_0)^2} = (P_0 + P) + \frac{\nu_\infty}{\lambda}(P_0 + P)^2$$
$$= (P_0 + P)\left\{1 + \frac{\nu_\infty}{\lambda}(P_0 + P)\right\}. \tag{5.30}$$

5.3 Extensions of Sect. 5.2

In this section, the Tait method will be extended [1–3] to polar molecules and liquid mixtures.

A form of Tait's equation to be utilized in this section [*Murnaghan–Tait*] is

$$\frac{\rho}{\rho_0} = \frac{1}{\left\{1 - C\ln\left|\frac{P+B}{P_0+B}\right|\right\}}, \tag{5.31}$$

where ρ is the density at pressure P, ρ_0 is the density at the reference pressure P_0, and $\{B, C\}$ are fitted parameters.

The isothermal compressibility at pressure P in terms of Eq. 5.31 is

$$\beta_{\mathrm{T}} = \frac{1}{\rho}\left(\frac{\partial\rho}{\partial P}\right)_{\mathrm{T}} = \frac{1}{\left\{1 - C\ln\left|\frac{P+B}{P_0+B}\right|\right\}}\frac{C}{P+B}. \tag{5.32}$$

And the isothermal compressibility at the reference pressure P_0 is

$$\beta_T = \frac{1}{\rho}\left(\frac{\partial\rho}{\partial P}\right)_T = \frac{C}{P+B}. \tag{5.33}$$

Solving Eq. 5.33 for C allows us to define Eqs. 5.31 and 5.32 in terms of one parameter, which is B. It will be shown that this B can be determined utilizing the idea of corresponding states, which is the basis to the Lee–Kessler chart [5].

Much as we can define a reduced temperature (T_{r}) and reduced pressure (P_{r}), we can also define a reduced B_{r} defined as

$$B_{\mathrm{r}} = \frac{B}{P_{\mathrm{c}}} = f(T_{\mathrm{r}}, \omega), \tag{5.34}$$

where

$$T_{\mathrm{r}} = \frac{T}{T_{\mathrm{c}}}, P_{\mathrm{r}} = \frac{P}{P_{\mathrm{c}}}. \tag{5.35}$$

And $\{T_{\mathrm{c}}, P_{\mathrm{c}}\}$ are the temperature and pressure at the critical point [6] and ω is the accentricity of the substance.

In Hoang et al. [5], it has been shown that

$$B_{\mathrm{r}} = -2 + \left[1 + k\left\{1 - \sqrt{T_{\mathrm{r}}}\right\}\right]^2. \tag{5.36}$$

And

$$k = -0.883\omega^2 + 14.34\omega + 10.17, \tag{5.37}$$

where ω is the accentricity discussed in Chap. 3.

The application of Tait's equation to liquid mixtures can be accomplished via Kay's mixing rules [5], which are

$$T_c^{mix} = \sum \chi_i T_{c,i} \tag{5.38}$$

$$P_c^{mix} = \sum \chi_i P_{c,i}. \tag{5.39}$$

And

$$\omega_{mix} = \sum \chi_i \omega_i, \tag{5.40}$$

where χ_i is the molar concentration of component i.

Example 5.4 will illustrate this process for methane.

Example 5.4 Determination of density and beta for methane from 0 to 50 MPa at 191 K (Table 5.3)

Table 5.3 Liquid mixture parameters for EoS [7]

Substance		Methane
T[c]	[K]	191
P[c]	[MPa]	4.5992
T	[K]	110
P	[MPa]	10
T[r]		0.6
P[r]		2.2
Omega		0.01
k(omega)		10.33
B[r]	[MPa]	10.19
B	[MPa]	46.87
P(0)	[Mpa]	10.00
Beta(T, 0)	[MPa]	1.81E – 03
Rho(T, 0)	[mole/l]	26.999
P(low)	[MPa]	0
P(high)	[MPa]	100
N		100
dP	[MPa]	1
MW	[kg/kmole]	16.042

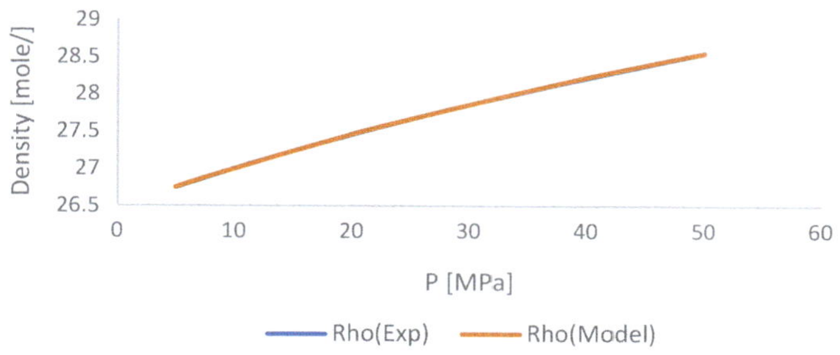

Fig. 5.5 Density versus *P*

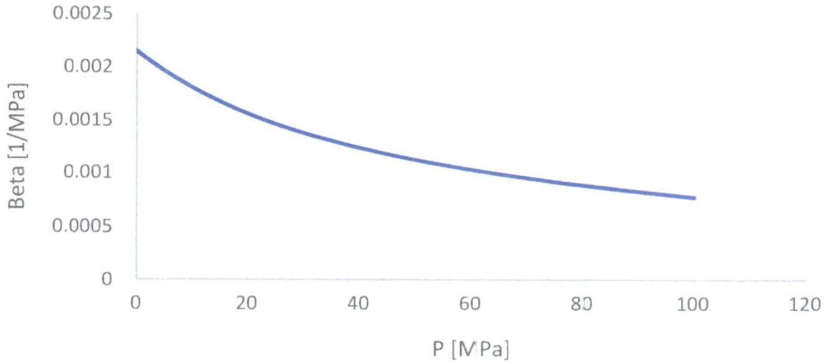

Fig. 5.6 Beta versus *P*

The density versus pressure for the model and experimental data is given as Fig. 5.5 and beta versus pressure is given as Fig. 5.6 where it is seen that the model and experimental data for density versus pressure is spot on.

5.4 Speed of Sound in Liquids

In this section, the speed of sound for fresh water, salt water, liquid mixtures, and polar molecules [9] will be explored.

5.4.1 Speed of Sound of Various Liquids Under Pressure

The speed of sound for a liquid can be determined in at least two ways. The two ways given here are

1. Derivation from the MT Equation of State.
2. From bulk modulus and density data.

Each method is given below.

Speed of Sound from MT EoS

Given the following model for liquid equations of state

$$\frac{P}{P_0} \approx (B+1)\left(\frac{\rho}{\rho_0}\right)^n - B. \tag{5.41}$$

And by definition of the speed of sound for a one phase (component) fluid is

$$c^2 \equiv \left(\frac{\partial P}{\partial \rho}\right)_s = \gamma\left(\frac{\partial P}{\partial \rho}\right)_T. \tag{5.42}$$

Applying this definition to Eq. 5.41 results in

$$c^2 = \gamma\left(\frac{\partial P}{\partial \rho}\right)_T = \gamma n(B+1)\frac{P_0}{\rho_0}\left(\frac{\rho}{\rho_0}\right)^{n-1}. \tag{5.43}$$

Speed of Sound from Bulk Modulus and Density

Another means to determining the speed of sound for a single phase (component) fluid is to determine the isentropic bulk modulus and density data for the substance where

$$c^2 \approx \frac{K_s}{\rho}. \tag{5.44}$$

Example 5.5 Speed of sound in fresh water under pressure

Determine the speed of sound through fresh water at 9000 atmospheres and 20 °C using the *Murnaghan–Tait* with B equal to 3000 and n equal to 7.

Where the solution is given below as $c = 2645$ m/s and compares well with the experimentally determined value of 2650 m/s [4].

Please note it was assumed $\gamma \approx 1$ (Fig. 5.7).

Example 5.6 Speed of sound through water

Determine the speed of sound through Mercury at standard temperature and pressure using the *Murnaghan–Tait* equation with the parameters estimated in Example 5.2 and compare against the established value of 1450 m/s [4].

Given that $\{n; B\} = \{6; 41,000\}$ from Example 5.4

$$c^2 \approx n(B+1)\frac{P_0}{\rho_0} = 6(41,001)\frac{101,300}{13,545}.$$

B	3000	
n	7	
P(a)	[atm]	1
Rho(a)	[kg/m^3]	998
P/P(a)		9000

$$\frac{P}{P_a} + B = (B+1)\left(\frac{\rho}{\rho_a}\right)^n \qquad \frac{dP}{d\rho} = n(B+1)\frac{P_a}{\rho_a}\left(\frac{\rho}{\rho_a}\right)^{n-1}$$

Rho	Rho/Rho(a)	LHS	RHS	Diff	dP/d(rho)	C
[kg/m^3]						[m/s]
1216.52	1.22	12000	12000.00	-3E-05	6994719	2645

Fig. 5.7 Solution for Example 5.5

And

$$c = 1356\ \frac{\text{m}}{\text{s}},$$

which is about 6% below the accepted value of 1450 m/s [4].

Again, please note it was assumed $\gamma \approx 1$.

5.4.2 *Speed of Sound in Liquid Mixtures*

The speed of sound for a liquid mixture is determined by utilizing the definition of beta given from Sect. 5.3.

$$\beta_{\text{T}} = \frac{1}{\rho}\left(\frac{\partial \rho}{\partial P}\right)_{\text{T}}. \tag{5.45}$$

And solving this equation for the definition of speed of sound

$$c^2 = \left(\frac{\partial P}{\partial \rho}\right)_{\text{s}} \approx \gamma\left(\frac{\partial P}{\partial \rho}\right)_{\text{T}} \approx \gamma \frac{1}{\beta_{\text{T}}}\frac{1}{\rho}. \tag{5.46}$$

One challenge with using Eq. 5.46 is that the gamma for a certain pressure, temperature and composition needs to be determined [7].

Example 5.6 Speed of sound of methane mixture

In Example 5.4, a model for density versus pressure for methane, which is a mixture, was shown to work when compared against data provided by the National Institute of Standards and Testing [7]. The same data set was compared against a model for speed of sound given as Eq. 5.46 (see Fig. 5.8) where the discrepancy is due to the

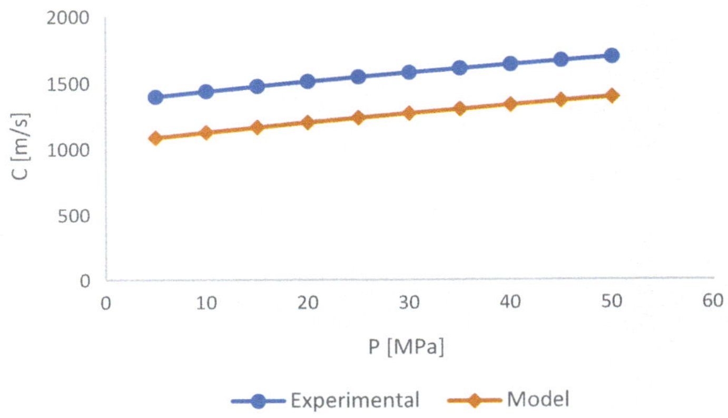

Fig. 5.8 Speed of sound versus pressure for methane

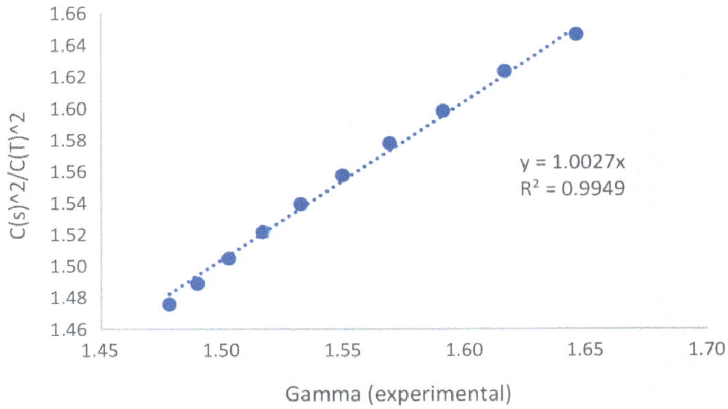

Fig. 5.9 $c_{\text{exp}}^2/c_{\text{model}}^2$ versus γ

fact that the experimental results are for an isentropic flow and the model provides results for an isothermal flow. Again, defining γ as

$$\gamma = \frac{c_{\text{s}}^2}{c_{\text{T}}^2}. \tag{5.47}$$

It is shown in Fig. 5.9 that the discrepancy in these two forms of speed of sound is exactly γ.

As previously noted, the model works well but does not include an estimate for γ, which would be required to get speed of sound along an isentropic path.

References

1. Dymond, Malhotra, The Tait equation: 100 years on. Int. J. Thermophys. **9**(6), 941–951 (1988)
2. Poling, Prausnitz, O'Connell, *The Properties of Gases and Liquids*, 5th edn. (McGraw-Hill, New York, NY)
3. Tait Equation. https://en.wikipedia.org/wiki/Tait_equation. Accessed 22 May 2023.
4. F.M. White, *Fluid Mechanics*, 4th edn. (Mc-Graw Hill, 1999)
5. Hoang, Galliero, Montel, Bickert, Tait Equation in the extended corresponding states framework: application to liquids and liquid mixtures. Fluid Phase Equilibria **387**, 5–11 (2015)
6. I.K. Gamwo, et al. High temperature, high pressure equation of state: solidification of hydrocarbons and measurement of Krytox oil using rolling-ball viscometer validation. National Energy Technology Laboratory, NETL-TRS-5-2014, 3 Oct 2014
7. NIST Webbook for Thermodynamics Properties of Fluids. https://webbook.nist gov/chemistry/fluid. Accessed 22 May 2023
8. Nayar, Sharqawy, Lienhard. http://web.mit.edu/seawater. Accessed 22 May 2023
9. Outcalt, Fortin, Density and speed of sound measurements of four bioderived aviation fuels. J. Chem. Eng. Data **57**, 2869–2877 (2012)

Chapter 6
Wave Speed in Solids

Abstract In this chapter, an approximate model is developed for the speed of sound through solids, which is only appropriate for a one-dimensional solid such as a long, thin metal rod. A wave equation is developed for a flexible wire and two possible scenarios presented—irrotational waves and solenoidal waves. These two scenarios provide the speed of sound associated with longitudinal and transverse waves that are also called primary and secondary waves and can be related to speed of sound of earthquakes. Comparisons are made between experimental data and the developed models for speed of sound through one-dimensional solids, longitudinal (primary) waves and transverse (secondary) waves. For the experimental data for several metals, the worst relative errors are—primary waves (5.2%), secondary waves (9.3%), and one-dimensional solid (6.8%).

Keywords Speed of sound (lateral waves) · Speed of sound (transverse waves) · Speed of sound through a one-dimensional solid · Young's modulus · Modulus of rigidity

6.1 Preview

We saw in Sect. 5.2 that the isentropic Bulk Modulus (K_s) is defined as

$$K_s = -\nu\left(\frac{\partial P}{\partial \nu}\right)_s = \rho\left(\frac{\partial P}{\partial \rho}\right)_s \tag{6.1}$$

And the Bulk Modulus is related to the Young's Modulus through

$$\frac{E}{K_s} = 3(1 - 2\sigma) \tag{6.2}$$

Where for most metals $E \approx K_s$, thus

H. Foust, *Speed of Sound and Critical Flow*, Shock Wave and High Pressure Phenomena, https://doi.org/10.1007/978-981-97-5918-7_6

$$E \approx \rho\left(\frac{\partial P}{\partial \rho}\right)_s \tag{6.3}$$

And the speed of sound through solids (for compression waves) is roughly

$$c^2 = \left(\frac{\partial P}{\partial \rho}\right)_s \approx \frac{E}{\rho} \tag{6.4}$$

I say roughly because this estimate of speed of sound is constrained to a one-dimensional solid such as a long, thin metal rod. Below we'll see that the speed of sound through solids depends on the direction of propagation relative to the direction of motion where in a three-dimensional solid there can be longitudinal waves and transverse waves as discussed in Chap. 1.

6.2 Wave Speed in Flexible Wires

Wave Speed for Flexible Wire

Given Newton's Second Law (and utilizing Fig. 6.1)

$$\vec{F} = m \cdot \vec{a}, \tag{6.5}$$

where $\vec{F}$ equals

$$\vec{F} = (-\vec{\tau}) + \left(\vec{\tau} + \frac{\partial \vec{\tau}}{\partial x}\mathrm{d}x\right) = \frac{\partial \vec{\tau}}{\partial x}\mathrm{d}x \tag{6.6}$$

dm equals

$$\mathrm{d}m = \lambda_0 \cdot \mathrm{d}x \tag{6.7}$$

And $\vec{a}$ equals

$$\vec{a} = \ddot{\vec{\tau}} = \frac{\partial^2 \vec{\rho}}{\partial t^2} \tag{6.8}$$

Substitution of Eqs. 6.6, 6.7, and 6.8 into Eq. 6.5 results in

$$\left(\frac{\partial \vec{\tau}}{\partial x}\mathrm{d}x\right) = (\lambda_0 \cdot \mathrm{d}x)\left(\frac{\partial^2 \vec{\rho}}{\partial t^2}\right) \tag{6.9}$$

Or

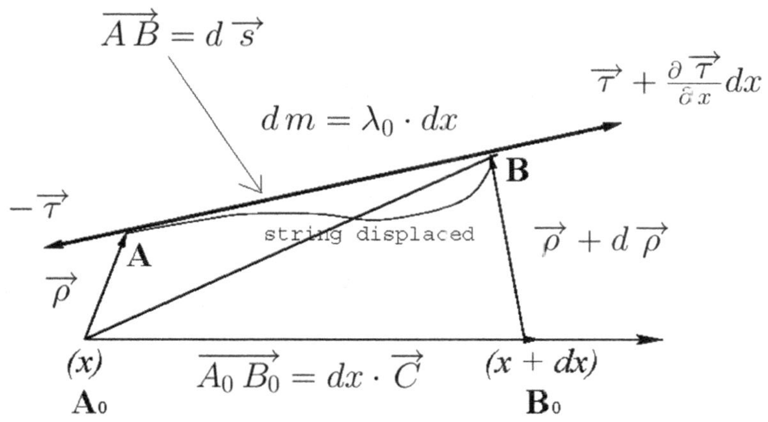

Fig. 6.1 Displaced string

$$\frac{\partial \vec{\tau}}{\partial x} = \lambda_0 \left(\frac{\partial^2 \vec{\rho}}{\partial t^2} \right) \tag{6.10}$$

In component form, $\vec{\rho}$ equals

$$\vec{\rho} = \varepsilon \cdot \hat{i} + \eta \cdot \hat{j} + \zeta \cdot \hat{k} \tag{6.11}$$

And, $\overrightarrow{d\rho}$ equals

$$\overrightarrow{\mathrm{d}\rho} = \mathrm{d}\varepsilon \cdot \hat{i} + \mathrm{d}\eta \cdot \hat{j} + \mathrm{d}\zeta \cdot \hat{k} \tag{6.12}$$

Four vectors are related,

$$\overrightarrow{A_0B_0} = \mathrm{d}x \cdot \hat{i}, \vec{\rho}, \vec{\rho} + \mathrm{d}\vec{\rho}, \text{ and } \overrightarrow{AB} = \mathrm{d}\vec{s} \tag{6.13a, b, c}$$

Because

$$\vec{c} = \vec{\rho} + \overrightarrow{AB} = \overrightarrow{A_0B_0} + (\vec{\rho} + \mathrm{d}\vec{\rho}) \tag{6.14}$$

And

$$\mathrm{d}\vec{s} = \overrightarrow{AB} = \overrightarrow{A_0B_0} + \mathrm{d}\vec{\rho} = \mathrm{d}x \cdot \hat{i} + \left(\mathrm{d}\varepsilon \cdot \hat{i} + \mathrm{d}\eta \cdot \hat{j} + \mathrm{d}\zeta \cdot \hat{k}\right) \tag{6.15}$$

Or

$$\mathrm{d}\vec{s} = \mathrm{d}x \cdot \left[\hat{i} + \frac{\partial \varepsilon}{\partial x} \cdot \hat{i} + \frac{\partial \eta}{\partial x} \cdot \hat{j} + \frac{\partial \zeta}{\partial x} \cdot \hat{k}\right] = \mathrm{d}x \cdot \left[\left(1 + \frac{\partial \varepsilon}{\partial x}\right) \cdot \hat{i} + \frac{\partial \eta}{\partial x} \cdot \hat{j} + \frac{\partial \zeta}{\partial x} \cdot \hat{k}\right] \tag{6.16}$$

And the magnitude of $|d\vec{s}|$ is

$$ds^2 = \left(1 + \frac{\partial \varepsilon}{\partial x}\right)^2 + \left(\frac{\partial \eta}{\partial x}\right)^2 + \left(\frac{\partial \zeta}{\partial x}\right)^2 \tag{6.17}$$

From the definition of Young's Modulus,

$$\sigma = E \cdot \varepsilon \tag{6.18}$$

where sigma is stress, E is Young's Modulus and epsilon is strain defined as

$$\varepsilon = \frac{l - l_0}{l_0} = \frac{\Delta l}{l_0} = \frac{ds - dx}{dx} = \frac{ds}{dx} - 1 \tag{6.19}$$

Such that

$$\frac{\Delta|\vec{\tau}|}{s} = E \cdot \left(\frac{ds}{dx} - 1\right) \tag{6.20}$$

And

$$|\vec{\tau}| = \tau_0 + EA \cdot \left(\frac{ds}{dx} - 1\right) \tag{6.21}$$

where A is the string's cross-sectional area

Due to the fact that the string is now flexible,

$$\vec{\tau} = |\vec{\tau}| \cdot \left(\frac{d\vec{s}}{ds}\right) = \left[\tau_0 + EA \cdot \left(\frac{ds}{dx} - 1\right)\right] \cdot \left(\frac{d\vec{s}}{ds}\right) \tag{6.22}$$

Assuming strains in three principal directions are very small, e.g.,

$$\left(\frac{\partial \varepsilon}{\partial x}\right) \ll 1, \left(\frac{\partial \eta}{\partial x}\right) \ll 1, \left(\frac{\partial \zeta}{\partial x}\right) \ll 1 \tag{6.23a, b, c}$$

Always us to state

$$\left(\frac{ds}{dx}\right)^2 = 1 + 2\left(\frac{\partial \varepsilon}{\partial x}\right) + \left(\frac{\partial \varepsilon}{\partial x}\right)^2 + \left(\frac{\partial \eta}{\partial x}\right)^2 + \left(\frac{\partial \zeta}{\partial x}\right)^2 = 1 + \left(2\frac{\partial \varepsilon}{\partial x} + 2\Delta\right), \tag{6.24}$$

where

$$\Delta = \frac{1}{2}\left[\left(\frac{\partial \varepsilon}{\partial x}\right)^2 + \left(\frac{\partial \eta}{\partial x}\right)^2 + \left(\frac{\partial \zeta}{\partial x}\right)^2\right] \tag{6.25}$$

Utilizing the Taylor series expansion for $\sqrt{1+u} = 1 + \frac{u}{2} - \frac{u^2}{8}$ one can get to

$$\frac{\mathrm{d}s}{\mathrm{d}x} = 1 + \left(\frac{\partial\varepsilon}{\partial x} + \Delta\right) - \frac{1}{8}\left(2\frac{\partial\varepsilon}{\partial x} + 2\Delta\right)^2 + \text{H.O.T.} \tag{6.26}$$

And

$$\frac{\mathrm{d}s}{\mathrm{d}x} \approx 1 + \frac{\partial\varepsilon}{\partial x} + \Delta - \frac{1}{2}\left(\frac{\partial\varepsilon}{\partial x}\right)^2 \tag{6.27}$$

And substituting for Δ the right-hand side of Eq. 6.25 results in

$$\frac{\mathrm{d}s}{\mathrm{d}x} \approx 1 + \frac{\partial\varepsilon}{\partial x} + \frac{1}{2}\left[\left(\frac{\partial\eta}{\partial x}\right)^2 + \left(\frac{\partial\zeta}{\partial x}\right)^2\right] \tag{6.28}$$

The directional factor $\frac{\mathrm{d}\vec{s}}{\mathrm{d}s}$ can be represented as

$$\frac{\mathrm{d}\vec{s}}{\mathrm{d}s} = \frac{\mathrm{d}\vec{s}}{\mathrm{d}x}\frac{\mathrm{d}x}{\mathrm{d}x} = \left(\frac{\mathrm{d}s}{\mathrm{d}x}\right)^{-1}\frac{\mathrm{d}\vec{s}}{\mathrm{d}x} \tag{6.29}$$

where

$$\frac{\mathrm{d}\vec{s}}{\mathrm{d}x} = \left(1 + \frac{\partial\varepsilon}{\partial x}\right)\cdot\hat{\imath} + \frac{\partial\eta}{\partial x}\cdot\hat{\jmath} + \frac{\partial\zeta}{\partial x}\cdot\hat{k} \tag{6.30}$$

s.t.

$$\frac{\mathrm{d}\vec{s}}{\mathrm{d}x} = \frac{1}{1 + \frac{\partial\varepsilon}{\partial x} + \frac{1}{2}\left[\left(\frac{\partial\eta}{\partial x}\right)^2 + \left(\frac{\partial\zeta}{\partial x}\right)^2\right]}\left\{\left(1 + \frac{\partial\varepsilon}{\partial x}\right)\cdot\hat{\imath} - \frac{\partial\eta}{\partial x}\cdot\hat{\jmath} + \frac{\partial\zeta}{\partial x}\cdot\hat{k}\right\} \tag{6.31}$$

where $1/\frac{\mathrm{d}s}{\mathrm{d}x}$ is of the form $\frac{1}{1+u}$ with Taylor Series

$$\frac{1}{1+u} = 1 - u + u^2 + \cdots \text{H.O.T.} \tag{6.32}$$

And

$$\frac{\mathrm{d}\vec{s}}{\mathrm{d}x} = \left\{1 - \left[\frac{\partial\varepsilon}{\partial x} + \frac{1}{2}\left\{\left(\frac{\partial\eta}{\partial x}\right)^2 + \left(\frac{\partial\zeta}{\partial x}\right)^2\right\}\right]\right\}\left\{\left(1 + \frac{\partial\varepsilon}{\partial x}\right)\cdot\hat{\imath} + \frac{\partial\eta}{\partial x}\cdot\hat{\jmath} + \frac{\partial\zeta}{\partial x}\cdot\hat{k}\right\} \tag{6.33}$$

or

$$\frac{\mathrm{d}\vec{s}}{\mathrm{d}x} = \left(1+\frac{\partial\varepsilon}{\partial x}\right)\cdot\hat{i} + \frac{\partial\eta}{\partial x}\cdot\hat{j} + \frac{\partial\zeta}{\partial x}\cdot\hat{k}$$

$$-\left\{\frac{\partial\varepsilon}{\partial x} + \frac{1}{2}\left\{\left(\frac{\partial\eta}{\partial x}\right)^2 + \left(\frac{\partial\zeta}{\partial x}\right)^2\right\}\right\}\left\{\left(1+\frac{\partial\varepsilon}{\partial x}\right)\cdot\hat{i} + \frac{\partial\eta}{\partial x}\cdot\hat{j} + \frac{\partial\zeta}{\partial x}\cdot\hat{k}\right\}$$

$$+\left\{\frac{\partial\varepsilon}{\partial x}\right\}^2\left\{\left(1+\frac{\partial\varepsilon}{\partial x}\right)\cdot\hat{i} + \frac{\partial\eta}{\partial x}\cdot\hat{j} + \frac{\partial\zeta}{\partial x}\cdot\hat{k}\right\}$$

$$= \left(1+\frac{\partial\varepsilon}{\partial x}\right)\cdot\hat{i} + \frac{\partial\eta}{\partial x}\cdot\hat{j} + \frac{\partial\zeta}{\partial x}\cdot\hat{k} - \frac{\partial\varepsilon}{\partial x}\left(1+\frac{\partial\varepsilon}{\partial x}\right)\cdot\hat{i}$$

$$-\frac{\partial\varepsilon}{\partial x}\frac{\partial\eta}{\partial x}\cdot\hat{j} - \frac{\partial\varepsilon}{\partial x}\frac{\partial\zeta}{\partial x}\cdot\hat{k}$$

$$-\frac{1}{2}\left\{\left(\frac{\partial\eta}{\partial x}\right)^2 + \left(\frac{\partial\zeta}{\partial x}\right)^2\right\}\left\{\left(1+\frac{\partial\varepsilon}{\partial x}\right)\cdot\hat{i}\right\}$$

$$-\frac{1}{2}\left\{\left(\frac{\partial\eta}{\partial x}\right)^2 + \left(\frac{\partial\zeta}{\partial x}\right)^2\right\}\left\{\frac{\partial\eta}{\partial x}\cdot\hat{j}\right\}$$

$$-\frac{1}{2}\left\{\left(\frac{\partial\eta}{\partial x}\right)^2 + \left(\frac{\partial\zeta}{\partial x}\right)^2\right\}\left\{\frac{\partial\zeta}{\partial x}\cdot\hat{k}\right\} + \left\{\frac{\partial\varepsilon}{\partial x}\right\}^2\left\{\left(1+\frac{\partial\varepsilon}{\partial x}\right)\cdot\hat{i}\right\}$$

$$+\left\{\frac{\partial\varepsilon}{\partial x}\right\}^2\left\{\frac{\partial\eta}{\partial x}\cdot\hat{j}\right\} + \left\{\frac{\partial\varepsilon}{\partial x}\right\}^2\left\{\frac{\partial\zeta}{\partial x}\cdot\hat{k}\right\} \tag{6.34}$$

$$\frac{\mathrm{d}\vec{s}}{\mathrm{d}x} = \left(1+\frac{\partial\varepsilon}{\partial x}\right)\cdot\hat{i} + \frac{\partial\eta}{\partial x}\cdot\hat{j} + \frac{\partial\zeta}{\partial x}\cdot\hat{k} - \frac{\partial\varepsilon}{\partial x}\cdot\hat{i} - \left\{\frac{\partial\varepsilon}{\partial x}\right\}^2\cdot\hat{i}$$

$$-\frac{\partial\varepsilon}{\partial x}\frac{\partial\eta}{\partial x}\cdot\hat{j} + \frac{\partial\varepsilon}{\partial x}\frac{\partial\zeta}{\partial x}\hat{k} - \frac{1}{2}\left[\left(\frac{\partial\eta}{\partial x}\right)^2 + \left(\frac{\partial\zeta}{\partial x}\right)^2\right]\cdot\hat{i} + \left(\frac{\partial\varepsilon}{\partial x}\right)^2\cdot\hat{i}$$

$$+\left(\frac{\partial\eta}{\partial x}\right)^2\cdot\hat{j} + \left(\frac{\partial\zeta}{\partial x}\right)^2\cdot\hat{k} \tag{6.35}$$

Equation 6.35 equals

$$\frac{\mathrm{d}\vec{s}}{\mathrm{d}x} = \hat{i}\cdot\left[1 - \frac{1}{2}\left(\frac{\partial\eta}{\partial x}\right)^2 - \frac{1}{2}\left(\frac{\partial\zeta}{\partial x}\right)^2\right] + \hat{j}\cdot\left[\frac{\partial\eta}{\partial x} - \frac{\partial\varepsilon}{\partial x}\frac{\partial\eta}{\partial x} + \left(\frac{\partial\varepsilon}{\partial x}\right)^2\right]$$

$$+\hat{k}\cdot\left[\frac{\partial\zeta}{\partial x} - \frac{\partial\varepsilon}{\partial x}\frac{\partial\zeta}{\partial x} + \left(\frac{\partial\varepsilon}{\partial x}\right)^2\right], \tag{6.36}$$

where $\left(\frac{\partial\varepsilon}{\partial x}\right)^2 \approx 0$.

Given

$$\begin{aligned}
\vec{\tau} &= \left[\tau_0 + ES \cdot \left(\frac{ds}{dx} - 1\right)\right] \cdot \frac{d\vec{s}}{ds} \\
&= \left[\tau_0 + ES \cdot \left\{\frac{\partial \varepsilon}{\partial x} + \frac{1}{2}\left[\left(\frac{\partial \eta}{\partial x}\right)^2 + \left(\frac{\partial \zeta}{\partial x}\right)^2\right]\right\}\right] \\
&\quad \cdot \left\{\hat{i} \cdot \left[1 - \frac{1}{2}\left(\frac{\partial \eta}{\partial x}\right)^2 - \frac{1}{2}\left(\frac{\partial \zeta}{\partial x}\right)^2\right] + \hat{j} \cdot \left[\frac{\partial \eta}{\partial x} - \frac{\partial \varepsilon}{\partial x}\frac{\partial \eta}{\partial x}\right] + \hat{k} \cdot \left[\frac{\partial \zeta}{\partial x} - \frac{\partial \varepsilon}{\partial x}\frac{\partial \zeta}{\partial x}\right]\right\} \\
&= \left[\tau_0 + ES\frac{\partial \varepsilon}{\partial x}\right] \cdot \left\{\hat{i} + \hat{j}\left(\frac{\partial \eta}{\partial x}\right) + \hat{k}\left(\frac{\partial \zeta}{\partial x}\right)\right\} \\
&= \hat{i} \cdot \left(\tau_0 + ES\frac{\partial \varepsilon}{\partial x}\right) + \hat{j}\tau_0\frac{\partial \eta}{\partial x} + \hat{k}\tau_0\frac{\partial \zeta}{\partial x}
\end{aligned} \tag{6.37}$$

Inserting the last equation into

$$\frac{\partial \vec{\tau}}{\partial x} = \lambda_0 \frac{\partial^2 \vec{\rho}}{\partial t^2} \tag{6.38}$$

Results in

$$\hat{i} \cdot \left(ES\frac{\partial^2 \varepsilon}{\partial x^2}\right) + \hat{j} \cdot \left(\frac{\partial^2 \eta}{\partial x^2}\tau_0\right) + \hat{k}\tau_0\frac{\partial^2 \zeta}{\partial x^2} = \lambda_0\left[\hat{i} \cdot \frac{\partial^2 \varepsilon}{\partial t^2}\right] + \lambda_0\left[\hat{j} \cdot \frac{\partial^2 \eta}{\partial t^2}\right] + \lambda_0\left[\hat{k} \cdot \frac{\partial^2 \zeta}{\partial t^2}\right] \tag{6.39}$$

or along the principle axis

$$ES\frac{\partial^2 \varepsilon}{\partial x^2} = \lambda_0\left[\frac{\partial^2 \varepsilon}{\partial t^2}\right] \tag{6.40}$$

$$\frac{\partial^2 \eta}{\partial x^2}\tau_0 = \lambda_0\left[\frac{\partial^2 \eta}{\partial t^2}\right] \tag{6.41}$$

And

$$\tau_0\frac{\partial^2 \zeta}{\partial x^2} = \lambda_0\left[\frac{\partial^2 \zeta}{\partial t^2}\right] \tag{6.42}$$

Thus

$$c_B^2 = \frac{ES}{\lambda_0} = \frac{E}{\rho_0}, \quad c_T^2 = \frac{\tau_0}{\lambda_0}, \quad c_T^2 = \frac{\tau_0}{\lambda_0} \tag{6.43}$$

Equation 6.43 shows that in a solid the wave speed is dependent on direction and that the longitudinal and transverse wave speeds are different. In the next section, models for speed of sound longitudinal and transverse will be derived and related to material properties of the solid such as Young's modulus, Modulus of Rigidity, and Poisson's ratio [1].

6.3 Wave Speed in Solids (Metals)

We learn in mechanics of materials that the strain of elastic materials under stress in three principal directions can be expressed as

$$\begin{bmatrix} \varepsilon_{xx} \\ \varepsilon_{yy} \\ \varepsilon_{zz} \end{bmatrix} = \frac{1}{E} \begin{bmatrix} 1 & -\upsilon & -\upsilon \\ -\upsilon & 1 & -\upsilon \\ -\upsilon & -\upsilon & 1 \end{bmatrix} \begin{bmatrix} \sigma_{xx} \\ \sigma_{yy} \\ \sigma_{zz} \end{bmatrix} \tag{6.44}$$

where E is Young's modulus and ν is Poisson's ratio

Solving for the stresses results in

$$\begin{bmatrix} \sigma_{xx} \\ \sigma_{yy} \\ \sigma_{zz} \end{bmatrix} = \lambda \begin{bmatrix} 1 & 1 & 1 \\ 1 & 1 & 1 \\ 1 & 1 & 1 \end{bmatrix} \begin{bmatrix} \varepsilon_{xx} \\ \varepsilon_{yy} \\ \varepsilon_{zz} \end{bmatrix} + \mu \begin{bmatrix} \varepsilon_{xx} \\ \varepsilon_{yy} \\ \varepsilon_{zz} \end{bmatrix} \tag{6.45}$$

where

$$\lambda = \frac{\upsilon E}{(1 + \upsilon)(1 - 2\upsilon)} \tag{6.46}$$

And

$$\mu = \frac{E}{1 + \upsilon} \tag{6.47}$$

Which are known as Lame's constants. Equations 6.46 and 6.47 can be represented in terms of the modulus of rigidity (G) as

$$\mu = 2G \tag{6.48}$$

And

$$\lambda = \frac{2\upsilon}{1 - 2\upsilon} G \tag{6.49}$$

where Young's modulus ε and the modulus of rigidity (G) are related by

$$G = \frac{E}{2(1 + 2\upsilon)} \tag{6.50}$$

We have previously defined dilation (relative change in volume) as

$$\theta = \frac{\Delta V}{V} \tag{6.51}$$

And in terms of strains in the three principal directions

$$\begin{aligned}\theta &= \frac{\left(\Delta x + \frac{\partial \xi}{\partial x}\right)\left(\Delta y + \frac{\partial \xi}{\partial y}\right)\left(\Delta z + \frac{\partial \xi}{\partial z}\right) - \Delta x \Delta y \Delta z}{\Delta x \Delta y \Delta z} \\ &= \frac{\mathrm{d}\xi}{\mathrm{d}x} + \frac{\mathrm{d}\xi}{\mathrm{d}y} + \frac{\mathrm{d}\xi}{\mathrm{d}z} = \varepsilon_{xx} + \varepsilon_{yy} + \varepsilon_{zz}\end{aligned} \tag{6.52}$$

The dilation can be written compactly as

$$\theta = \nabla \cdot \underset{\sim}{\varrho} \tag{6.53}$$

Pressure can be defined in terms of dilation as

$$P = -K_s \theta \tag{6.54}$$

where K_s is the bulk modulus and related to Young's modulus (E) through

$$\theta = -3\frac{P}{E}(1 - 2\upsilon) \tag{6.55}$$

Or

$$E = 3K_s(1 - 2\upsilon) \tag{6.56}$$

where often for metals $\upsilon \approx 1/3$ and as such

$$E \approx K_s \tag{6.57}$$

Going back to Eq. 6.44

$$\begin{bmatrix} \varepsilon_{xx} \\ \varepsilon_{yy} \\ \varepsilon_{zz} \end{bmatrix} = \frac{1}{E}\begin{bmatrix} 1 & -\upsilon & -\upsilon \\ -\upsilon & 1 & -\upsilon \\ -\upsilon & -\upsilon & 1 \end{bmatrix}\begin{bmatrix} \sigma_{xx} \\ \sigma_{yy} \\ \sigma_{zz} \end{bmatrix} \tag{6.58}$$

In a more general frame of reference utilizing tensor notation, Eq. 6.58 can be expressed as

$$\mathbf{E} = \frac{1+\upsilon}{E}\boldsymbol{F} + \frac{3\upsilon}{E}\overline{\boldsymbol{P}}1, \tag{6.59}$$

where

$$\mathbf{E} = \begin{bmatrix} \varepsilon_{xx} \\ \varepsilon_{yy} \\ \varepsilon_{zz} \end{bmatrix}, \; \boldsymbol{F} = \lambda(\nabla \cdot \varrho)1 + 2\mu\mathbf{E}, \; \bar{\boldsymbol{p}} = -\frac{1}{3}\{\sigma_{xx} + \sigma_{yy} + \sigma_{zz}\} \tag{6.60}$$

From Newton's Second Law,

$$F = \nabla \cdot \boldsymbol{F} = \nabla \cdot \{\lambda(\nabla \cdot \varrho)1 + 2\mu\mathbf{E}\} = \rho_0 \frac{\partial^2 \varrho}{\partial t^2} \tag{6.61}$$

The divergence of the two $\boldsymbol{F}$ terms results in

$$(\lambda + \mu)\nabla\nabla \cdot \varrho + \mu\nabla \cdot \nabla\varrho = \rho_0 \frac{\partial^2 \varrho}{\partial t^2} \tag{6.62}$$

Two particular cases will be considered

1. Irrotational Waves ($\nabla \times \varrho = 0$)
2. Solenoidal Waves ($\nabla \cdot \varrho = 0$).

These two cases provide solutions for a longitudinal wave in an elastic solid and a transverse wave in an elastic solid [1].

Irrotational Waves

For an irrotational wave,

$$\nabla \times (\nabla \times \varrho) = \nabla\nabla \cdot \varrho - \nabla \cdot \nabla\varrho = 0 \tag{6.63}$$

Such that

$$\nabla\nabla \cdot \varrho = \nabla \cdot \nabla\varrho \tag{6.64}$$

And substituting into Eq. 6.62 results in

$$(\lambda + 2\mu)\nabla\nabla \cdot \varrho = \varrho_0 \frac{\partial^2 \varrho}{\partial t^2} \tag{6.65}$$

Or

$$\nabla\nabla \cdot \varrho = \frac{1}{c_l^2} \frac{\partial^2 \varrho}{\partial t^2} \tag{6.66}$$

where the wave speed, longitudinal is

$$c_l^2 = \frac{\lambda + 2\mu}{\rho_0} = \frac{E}{\rho_0} \frac{(1 - \upsilon)}{(1 + \upsilon)(1 - 2\upsilon)} \tag{6.67}$$

Solenoidal Waves

Given

$$(\lambda + \mu)\nabla\nabla \cdot \varrho + \mu\nabla \cdot \nabla\varrho = \rho_0 \frac{\partial^2 \varrho}{\partial t^2} \tag{6.68}$$

And assuming $\nabla \cdot \varrho = 0$ results in

$$\mu\nabla \cdot \nabla\varrho = \rho_0 \frac{\partial^2 \varrho}{\partial t^2} \tag{6.69}$$

Or

$$\nabla \cdot \nabla\varrho = \frac{1}{c_t^2} \frac{\partial^2 \varrho}{\partial t^2}, \tag{6.70}$$

where the wave speed, transverse is

$$c_t^2 = \frac{\mu}{\rho_0} = \frac{E}{\rho_0} \frac{1}{2(1 + \upsilon)} \tag{6.71}$$

As we have just shown Primary wave speed (longitudinal wave speed) is

$$c_p = \sqrt{\frac{E(1 - \nu)}{\rho(1 + \nu)(1 - 2\nu)}} \tag{6.72}$$

Secondary wave speed (transverse wave speed) is

$$c_s = \sqrt{\frac{G}{\rho}} \tag{6.73}$$

where

$$E = 2G(1 + \nu) \tag{6.74}$$

And for one-dimensional solids

$$c = \sqrt{\frac{E}{\rho}} \tag{6.75}$$

Given the following physical properties of selected metals (see Table 6.1).

Table 6.1 Physical properties of various metals [2]

Substance	E (GPa)	Poisson ratio	Density (gm/cm^3)	v(p) (m/s)	v(s) (m/s)	v(1D) (m/s)
Aluminum	70.3	0.345	2.7	6420	3040	5000
Copper (rolled)	129.8	0.343	8.96	5010	2270	3750
Copper (annealed)	129.8	0.345	8.96	4769	2324	3810
Nickel	219.2	0.306	8.9	6040	3000	4900
Tungsten (drawn)	411	0.28	19.3	5410	2640	4320
Tungsten (annealed)	411	0.28	19.3	5220	2890	4620
Titanium	115.7	0.321	4.506	6070	3125	5090

Where v(p) is the lateral wave speed (Primary Wave), v(s) is the longitudinal wave speed (Secondary Wave) and v(1D) is the wave speed in a one-dimensional solid (long, slender rod)

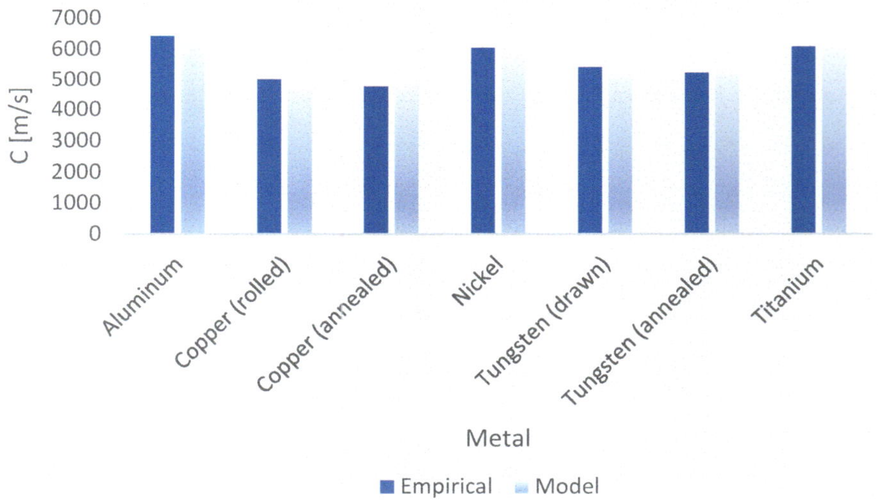

Fig. 6.2 P wave speeds (worst relative error is 5.2%)

Using the equations given above as Eqs. 6.73 through 6.75 and the physical properties of Table 6.1, the following comparisons were made between empirical and model results for primary waves, secondary waves, and waves in 1D solids. The empirical results were derived from [2] (Figs. 6.2, 6.3 and 6.4).

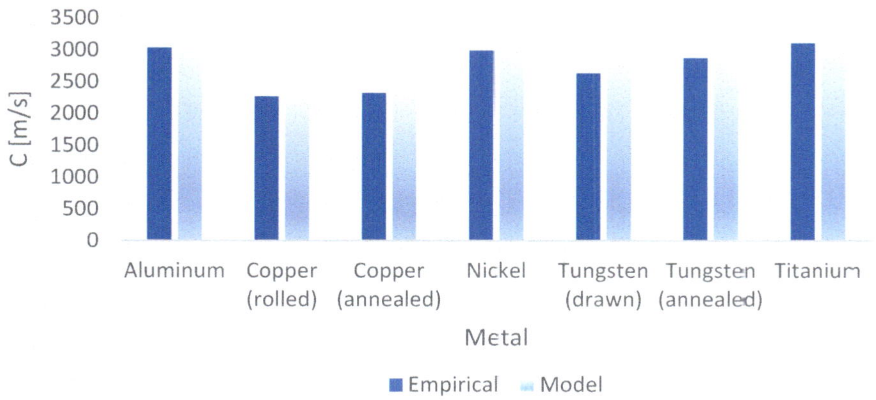

Fig. 6.3 S wave speeds (worst relative error is 9.3%)

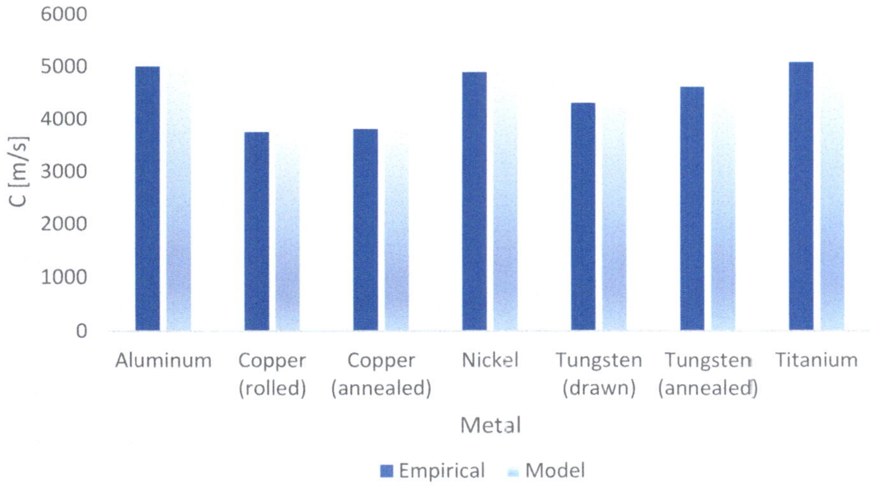

Fig. 6.4 Wave speed in 1D solids (worst relative error is 6.8%)

References

1. Elmore, Heald, *Physics of Waves* (Dover Books, 1985)
2. J.R. Rumble, *Handbook of Chemistry and Physics*, 102th edn. (CRC Press, Boca Raton, FL, 2021)

Chapter 7
Critical Flow, One Phase (Component)

Abstract The mass efflux associated with an ideal, perfect gas moving through a choke point is derived from six conditions/assumptions and equations for the first and second laws of thermodynamics; this model is then extended to a van der Waal gas. The last section demonstrates that often a liquid phase develops as the gas is depressurized and shows the possible errors of ignoring the development of this liquid phase for a fluid that is initially a pure gas.

Keywords Critical flow for an ideal · Perfect gas · Homogenous equilibrium flow · Non-equilibrium critical flow · Frozen equilibrium flow

7.1 Preview

The system under consideration is a pressurized vessel in which the discharge point is referred to as the "choke point," and the center of the tank is denoted the "stagnation point," which is depicted in Fig. 7.1.

The equation for the mass rate exiting (mass efflux) the system is

$$\frac{\mathrm{d}m(\mathrm{sys})}{\mathrm{d}t} = -A^* \rho^* u^* \tag{7.1}$$

where m is the mass of fluid within the system at time t, A^* is the choke cross-sectional area, ρ^* is the density at the choke point, and u^* is the local speed of sound at the choke point.

The crux of the problem is to develop a model for "$\rho^* u^*$"in terms of the properties within the tank, such as T_0 and P_0, where T_0 is the stagnation temperature and P_0 is the stagnation pressure.

Please note—Fundamentals of Fluid Mechanics by Munson et al [5] does an excellent job of explaining static and stagnant conditions in terms of pressures.

The development of a model for single-phase flow for $\rho^* u^*$ involved six assumptions/conditions that are

H. Foust, *Speed of Sound and Critical Flow*, Shock Wave and High Pressure Phenomena, https://doi.org/10.1007/978-981-97-5918-7_7

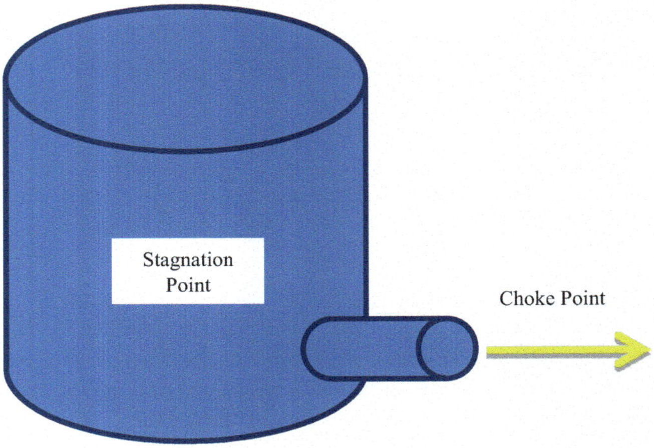

Fig. 7.1 Pressurized vessel

1. The reversibility of the system
2. An adiabatic system
3. A model for specific heat
4. The equation of state
5. The stagnation condition
6. Choked flow.

This chapter will introduce critical flow associated with a single phase (single component) and go on to discuss critical flow in a two phase (two component system). The concept of critical flow for a single-phase fluid can be understood in the following manner.

Given the following equation for 1st law.

$$0 = \Delta h + \Delta KE, \tag{7.2}$$

where the system has no work, no heat transfer, nor changes in potential energy, which has the differential version

$$0 = \mathrm{d}h + u\mathrm{d}u \tag{7.3}$$

And given the following equation for second law for an isentropic flow

$$T\mathrm{d}s = \mathrm{d}h - \frac{\mathrm{d}P}{\rho} = 0 \tag{7.4}$$

Equating the two forms of "dh" results in

$$-u\mathrm{d}u = \frac{\mathrm{d}P}{\rho} \tag{7.5}$$

or

$$\frac{\mathrm{d}u}{u} = -\frac{\mathrm{d}P}{\rho u^2} \tag{7.6}$$

And defining the mass flux (G) as

$$G = \rho u \tag{7.7}$$

or

$$\ln(G) = \ln(\rho) + \ln(u) \tag{7.8}$$

Taking the derivative of both sides results in

$$\frac{\mathrm{d}G}{G} = \frac{\mathrm{d}\rho}{\rho} + \frac{\mathrm{d}u}{u} \tag{7.9}$$

And Substituting Eq. 7.6 into Eq. 7.9 results in

$$\frac{\mathrm{d}G}{G} = -\frac{\mathrm{d}P}{\rho u^2} + \frac{\mathrm{d}P}{\rho}\left(\frac{\partial \rho}{\partial P}\right)_s \tag{7.10}$$

or

$$\frac{\mathrm{d}G}{G} = \rho \mathrm{d}P\left[-\frac{1}{G^2} + \frac{1}{\rho^2}\left(\frac{\partial \rho}{\partial P}\right)_s\right] \tag{7.11}$$

And
G is a maxima when

$$\left(\frac{\mathrm{d}G}{\mathrm{d}P}\right)_s = 0 \tag{7.12}$$

Such that

$$-\frac{1}{G^2} + \frac{1}{\rho^2}\left(\frac{\partial \rho}{\partial P}\right)_s = 0 \tag{7.13}$$

or

$$G_{\max} = \rho\sqrt{\left(\frac{\partial P}{\partial \rho}\right)_s} = \sqrt{\left(-\frac{\partial P}{\partial v}\right)_s} \tag{7.14}$$

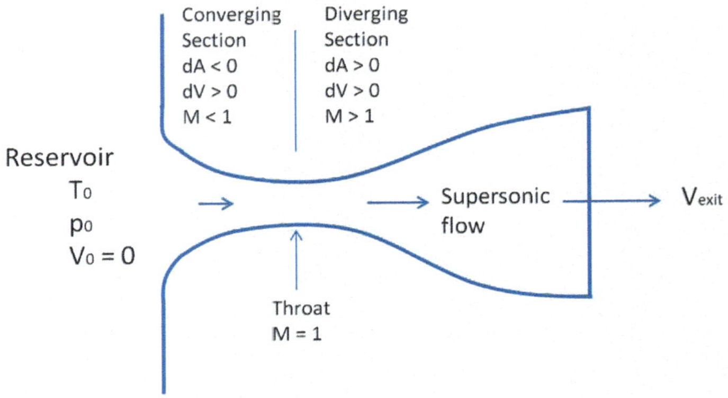

Fig. 7.2 Supersonic nozzle [1]

Additionally, the speed of sound for a single phase, single component substance is

$$a^2 = \left(\frac{\partial P}{\partial \rho}\right)_s \tag{7.15}$$

Such that we see that $\frac{G_{max}}{\rho}$ is equal to the local speed of sound!

Equation 7.15 is correct for a pure gas, but when the substance is a mixture of gas and liquid, the path is not necessarily isentropic.

We see in Fig. 7.2, which was shown original in Sect. 2.6 that the throat is the choke point and the Mach number is 1 consistent with the above discussion.

In Sect. 7.2, we will assume the gas is both ideal and thermodynamically perfect, e.g., specific heat is independent of temperature. With the six above assumptions/ conditions and an ideal, perfect gas, we get two algebraic expressions from the energy and entropy balances.

In Sect. 7.3, we will assume the gas is a van der Waal gas and that the specific heat is independent of temperature and pressure. Again, from the entropy and energy balances we get two algebraic expressions; how differently the two equations of state are in terms of critical flow will be explored through an example.

7.2 Ideal Gas

In this section, a critical flow model is developed for an ideal, perfect gas.

Entropy Balance

From the first four assumptions/conditions and the entropy balance, we get to the following relationship

$$ds = \frac{C_v dT}{T} + \frac{P dv}{T} \tag{7.16}$$

or

$$s_2 - s_1 = 0 = \frac{C_p}{\gamma} \int_{T_0}^{T^*} \frac{dT}{T} + \int_{v_0}^{v^*} \frac{R_u T}{vT} dv, \tag{7.17}$$

where γ is defined as

$$\gamma = \frac{c_p}{c_v} \tag{7.18}$$

And

$$\frac{c_P}{\gamma} \ln\left(\frac{T^*}{T_0}\right) + R_u \ln\left(\frac{v^*}{v_0}\right) = 0 \tag{7.19}$$

For an ideal gas

$$c_p - c_v = R_u \tag{7.20}$$

Substituting Eq. 7.20 into Eq. 7.19 results in

$$\frac{c_P}{\gamma} \ln\left(\frac{T^*}{T_0}\right) + (c_p - c_v) \ln\left(\frac{v^*}{v_0}\right) = 0 \tag{7.21}$$

And dividing by C_p gives

$$\frac{1}{\gamma} \ln\left(\frac{T^*}{T_0}\right) + \left(1 - \frac{1}{\gamma}\right) \ln\left(\frac{v^*}{v_0}\right) = 0 \tag{7.22}$$

or

$$\ln\left(\frac{T^*}{T_0}\right) + (\gamma - 1) \ln\left(\frac{v^*}{v_0}\right) = 0 \tag{7.23}$$

Thus

$$\frac{T^*}{T_0} = \left(\frac{v_0}{v^*}\right)^{\gamma - 1} \tag{7.24}$$

Energy Balance

From the last four assumptions/conditions and the energy balance, we get the following relationship

$$h_0 = h^* + \frac{1}{2}c^2, \tag{7.25}$$

where h_0 is the enthalpy at the stagnation point, h^* is the enthalpy at the discharge (choke) point, and c is the local speed of sound [1]. In a converging area, the chocked flow will be sonic when the following condition [1] is met

$$\frac{P^*}{P_0} = \left(\frac{2}{\gamma + 1}\right)^{\gamma/\gamma-1} \tag{7.26}$$

Which as a rule of thumb, essentially states that as long as the stagnation pressure is at least twice of the chocked pressure, then the discharge velocity is equal to the local speed of sound.

We saw in Chaps. 1 and 3 that c for an ideal, perfect gas is equal to

$$c = \sqrt{\gamma RT} \tag{7.27}$$

Substituting Eq. 7.27 into 7.25 for a perfect gas results in

$$c_p T_0 = c_p T^* + \frac{1}{2}\gamma RT^* \tag{7.28}$$

And dividing both sides by $c_p T^*$ results in

$$\frac{T_0}{T^*} = 1 + \frac{1}{2C_p}\gamma R \tag{7.29}$$

Further,

$$\frac{T_0}{T^*} = 1 + \frac{1}{2}\frac{1}{c_v}(c_p - c_v) \tag{7.30}$$

And

$$\frac{T_0}{T^*} = 1 + \frac{1}{2}(\gamma - 1) = \frac{1}{2}(\gamma + 1) \tag{7.31}$$

Again the mass rate of discharge is

$$\dot{m}_{sys} = -A^* u^* \rho^* \tag{7.32}$$

Given above with the definition for the local speed of sound (Eq. 3.21) for an ideal gas and

$$u^* = \sqrt{\gamma R_g T^*} \tag{7.33}$$

R(g)	[J/kg-K]	188.9
Gamma		1.289
T(0)	[K]	300
P(0)	[Pa]	2,000,000
h	[s]	0.1
V	[m^3]	1
A*	[m^2]	5.00E-03

Time	m(t)	P(t)	f[T0,P0]	m(t+1)
[s]	[kg]	[Pa]		[kg]
0	35.29	2000000	-27.946	32.50
0.1	32.50	1841630	-25.733	29.92
0.2	29.92	1695801	-23.695	27.55
0.3	27.55	1561519	-21.819	25.37
0.4	25.37	1437870	-20.091	23.36
0.5	23.36	1324013	-18.5	21.51
0.6	21.51	1219171	-17.035	19.81
0.7	19.81	1122631	-15.687	18.24
0.8	18.24	1033736	-14.444	16.80
0.9	16.80	951879	-13.301	15.47
1	15.47	876505	-12.247	14.24
1.1	14.24	807099	-11.278	13.11
1.2	13.11	743189	-10.385	12.08
1.3	12.08	684340	-9.5623	11.12
1.4	11.12	630150	-8.8051	10.24
1.5	10.24	580252	-8.1078	9.43
1.6	9.43	534305	-7.4658	8.68
1.7	8.68	491996	-6.8746	7.99
1.8	7.99	453037	-6.3303	7.36

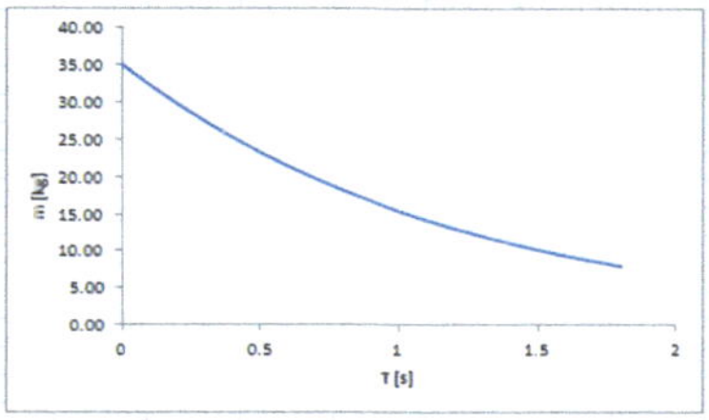

Fig. 7.3 Example 7.1 Results

Additionally, we know from Eqs. 7.24 and 7.31 that

$$\frac{\rho^*}{\rho_0} = \left(\frac{T^*}{T_0}\right)^{1/\gamma-1} = \left(\frac{2}{\gamma+1}\right)^{1/\gamma-1} \tag{7.34}$$

Substitution of Eqs. 7.33 and 7.34 (plus the ideal gas law) into 7.32 results in

$$\dot{m}_{\mathrm{sys}} = -A^*\sqrt{\frac{2R\gamma T_0}{\gamma+1}}\frac{P_0}{RT_0}\left(\frac{2}{\gamma+1}\right)^{1/\gamma-1} \tag{7.35}$$

Example 7.1 Pressure Vessel for Ideal Gas

Determine the amount of carbon dioxide within a pressure vessel after 1.5 s when the initial stagnation pressure is 2 MPa, temperature is 300 K and the opening is 5e − 3 m^2. The volume of the vessel is 1 m^3.

Note that CO_2 at this temperature and initial pressure is not ideal! In fact, using a Peng-Robinson Equation of State [1], it can be shown that the Z factor is 0.89, which tells us this gas is not acting ideal.

From the graph, the answer is about 10.24 kg (Fig. 7.3).

7.3 van der Waal Gas

In order to incorporate real gas effects, the following approach was adopted and includes the use of the van der Waal's equation of state

$$P = \frac{RT}{v-b} - \frac{a}{v^2}, \tag{7.36}$$

where $\{a, b\}$ have been previously defined.

Because the equation of state is more mathematically involved, the entropy and energy balance equation will be more involved.

Energy Balance

From the last four assumptions, an **energy balance** is given as

$$h_0 = h^* + \frac{1}{2}c^2, \tag{7.37}$$

where h_0 is the enthalpy at the stagnation point, h^* is the enthalpy at the discharge (choke) point, and c is the local speed of sound [2] for a van der Waal gas.

What we'll do is define "du", solve this equation and then define "dh" in terms of "du."

$$\mathrm{d}u = \left(\frac{\partial u}{\partial T}\right)_{\mathrm{v}} \mathrm{d}T + \left(\frac{\partial u}{\partial v}\right)_{\mathrm{T}} \mathrm{d}v, \tag{7.38}$$

where

$$\left(\frac{\partial u}{\partial T}\right)_{\mathrm{v}} = c_{\mathrm{v}} \tag{7.39}$$

And

$$\left(\frac{\partial u}{\partial v}\right)_{\mathrm{T}} = T\left(\frac{\partial P}{\partial T}\right)_{\mathrm{v}} - P = T\frac{R}{v-b} - \left\{\frac{R}{v-b} - \frac{a}{v^2}\right\} = \frac{a}{v^2} \tag{7.40}$$

So that

$$\mathrm{d}u = c_{\mathrm{v}}\mathrm{d}T + \frac{a}{v^2}\mathrm{d}v \tag{7.41}$$

And integrating

$$\Delta u = c_{\mathrm{v}}\Delta T + \frac{a}{v_1} - \frac{a}{v_2} \tag{7.42}$$

And

$$\begin{aligned}\Delta h = h_2 - h_1 &= \Delta u + P_2 v_2 - P_1 v_1 \\ &= c_{\mathrm{v}}\Delta T + \frac{a}{v_1} - \frac{a}{v_2} + \frac{RT_2 v_2}{v_2 - b} - \frac{a}{v_2} - \frac{RT_1 v_1}{v_1 - b} + \frac{a}{v_1}\end{aligned} \tag{7.43}$$

Additionally, the speed of sound for a van der Waal gas is [2]

$$c_{vw}^2 = \left(1 + \frac{R}{c_v}\right)\frac{v^2RT}{(v-b)^2} - 2\frac{a}{v} \tag{7.44}$$

And substituting Eqs. 7.43 and 7.44 into 7.37 results in

$$c_v\Delta T + \frac{a}{v_1} - \frac{a}{v_2} + \frac{RT_2v_2}{v_2 - b} - \frac{a}{v_2} - \frac{RT_1v_1}{v_1 - b} + \frac{a}{v_1} + \frac{1}{2}\left(1 + \frac{R}{c_v}\right)\frac{v_2^2RT_2}{(v_2 - b)^2} - \frac{a}{v_2} = 0 \tag{7.45}$$

Which can be slightly simplified into

$$c_v(T_2 - T_1) + 2\frac{a}{v_1} - 3\frac{a}{v_2} + R\left\{\frac{T_2v_2}{v_2 - b} - \frac{T_1v_1}{v_1 - b} + \frac{1}{2}\left(1 + \frac{R}{c_v}\right)\frac{v_2^2T_2}{(v_2 - b)^2}\right\} = 0 \tag{7.46}$$

Entropy Balance

From the first four assumptions, an entropy balance is given as

$$T\mathrm{d}s = \mathrm{d}u + P\mathrm{d}v \tag{7.47}$$

And

$$\mathrm{d}u = c_v\mathrm{d}T + \frac{a}{v^2}\mathrm{d}v \tag{7.48}$$

Substituting Eq. 7.48 into Eq. 7.47 results in

$$T\mathrm{d}s = c_v\mathrm{d}T + \left(P + \frac{a}{v^2}\right)\mathrm{d}v = c_v\mathrm{d}T + \frac{RT}{v - b}\mathrm{d}v \tag{7.49}$$

And

$$\mathrm{d}s = c_v\frac{\mathrm{d}T}{T} + \frac{R}{v - b}\mathrm{d}v \tag{7.50}$$

For isentropic flow

$$c_v \ln\left(\frac{T_2}{T_1}\right) + R\ln\left(\frac{v_2 - b}{v_1 - b}\right) = 0 \tag{7.51}$$

or

$$\ln\left(\frac{T_2}{T_1}\right) + (\gamma - 1)\ln\left(\frac{v_2 - b}{v_1 - b}\right) = 0 \tag{7.52}$$

And finally

$$\frac{T_2}{T_1} = \left(\frac{v_1 - b}{v_2 - b}\right)^{\gamma-1} \tag{7.53}$$

Critical flow could then be determined by solving Eqs. 7.46 and 7.53 to determine specific volume and temperature at the choked point based on the specific volume and temperature at the stagnation point.

7.4 Real Gas Effects

In this section, we begin to explore some of the limitations of assuming an ideal, perfect gas when determining critical flow for a system. We also explore the effects of a two-phase fluid on critical flow.

This discussion will be facilitated through an example.

Example 7.2 Real Gas Effects

Imagine a tank holds superheated vapor at 2000 kPa and 250 °C. This water expands to 101.3 kPa and 100 °C at the choke point, which will involve a phase change from superheated vapor to a mixture.

We'll see how this phase change and other real gas effects affect the critical flow calculated.

The simplest way to model critical flow for a two-phase system is as follows.

Given the following energy balance

$$h_i + \frac{u_i^2}{2} = h_e + \frac{u_e^2}{2}, \tag{7.54}$$

where now there are two phases and along with stagnation condition ($u_i \approx 0$) results in

$$h_i = h_e + \frac{u_e^2}{2} \tag{7.55}$$

And the conservation of mass for a two-phase system at the nozzle is

$$G^* = \frac{u_e}{v_{g,e}}, \tag{7.56}$$

where now the final specific volume and enthalpy are determined from knowing the quality, which is determined from assumption (1) given below.

Where the following assumptions are made

1. Flow is isentropic, $s_i = s_e$
2. substance goes to mixture
3. Assumed choke point is at ambient conditions.

This method for solving two-phase critical flow systems is known as the homogenous equilibrium model.

Given below are the solutions using the ideal, perfect gas critical flow model and homogenous equilibrium model (Figs. 7.4 and 7.5).

			Comments
State		Superheated Vapour	
P(0)	[kPa]	2,000	
T(0)	[C]	250	
nu(0)	[m^3/kg]	0.11144	
h(0)	[kJ/kg]	2902.46	
s(0)	[kJ/kg-K]	6.5452	
State		Mixture	
P(*)	[kPa]	101.3	
T(*)	[C]	100	
nu(l)	[m^3/kg]	0.001044	
nu(g)	[m^3/kg]	1.6729	
h(l)	[kJ/kg]	419.02	
h(g)	[kJ/kg]	2676.05	
s(l)	[Kj/kg-K]	1.3068	
s(g)	[Kj/kg-K]	7.3548	
R(g)	[kJ/kg-K]	0.4615	
Gamma		1.327	at 25 C, 100 kPa

x	f(x)
0	1.3068
1	7.3548

x	v0(x)	v1(x)	g(x)	Diff
86.61%	0.13386	0.86614	6.5452	0

x	h(l)	h(g)	h(mix)
	[kJ/kg]	[kJ/kg]	[kJ/kg]
36.61%	419.02	2676.05	2373.9

x	nu(l)	nu(g)	nu(mix)
	[m^3/kg]	[m^3/kg]	[m^3/kg]
36.61%	0.00104	1.6729	1.45

u*	LHS	RHS	Diff	G[c]
[m/s]	[J/kg]	[J/kg]		[kg/m^2-s]
1028.15	2902460	2902460	2.2E-07	709.51

Fig. 7.4 Homogenous equilibrium model

			Comments
State		Superheated Vapour	
P(0)	[kPa]	2,000	
T(0)	[C]	250	
nu(0)	[m^3/kg]	0.11144	
h(0)	[kJ/kg]	2902.46	
s(0)	[kJ/kg-K]	6.5452	
State		Mixture	
P(*)	[kPa]	101.3	
T(*)	[C]	100	
nu(l)	[m^3/kg]	0.001044	
nu(g)	[m^3/kg]	1.6729	
h(l)	[kJ/kg]	419.02	
h(g)	[kJ/kg]	2676.05	
s(l)	[Kj/kg-K]	1.3068	
s(g)	[Kj/kg-K]	7.3548	
R(g)	[kJ/kg-K]	0.4615	
Gamma		1.327	at 25 C, 100 kPa

rho(0)	[kg/m^3]	8.29
rho(*)	[kg/m^3]	5.21
T(0)	[K]	523
T(*)	[K]	450
C(*)	[m/s]	524.67
G[c]	[kg/m^2-s]	2736.1

$$\frac{\rho^*}{\rho_0} = \left(\frac{T^*}{T_0}\right)^{1/\gamma-1} = \left(\frac{2}{\gamma+1}\right)^{1/\gamma-1}$$

$$\frac{T_0}{T^*} = 1 + \frac{1}{2}(\gamma - 1) = \frac{1}{2}(\gamma + 1)$$

Fig. 7.5 Ideal, perfect gas model

From the ideal, perfect critical flow model it is determined that $G[c]$ is 2736 kg/m^2 s and from the homogenous equilibrium model it is determined that G[c] is 709.5 kg/m^2 s, which is drastically different. The homogenous equilibrium model itself ignore non-equilibrium effects such as mass transfer between phases, momentum transfer between phases, and heat transfer between phases. These non-equilibrium effects will be explored in the next chapter where we mainly focus on the work of Henry and Fauske [2–4].

Note to this point, we haven't discussed the importance of geometry on non-equilibrium effects, but we'll get back to geometry in Chap. 9.

References

1. H. Foust, *Thermodynamics, Gas Dynamics, and Combustion* (Springer, New York, NY, 2022)
2. H.K. Fauske, Contribution to the Theory of Two-Phase, One-Component Critical Flow. Argonne National Laboratory, ANL-6633 (1962)
3. R.E. Henry, A study of one- and two-component, two-phase critical flows at low qualities. Ph.D. Thesis, University of Notre Dame (1968)
4. R.E. Henry, H.K. Fauske, The two-phase critical flow of one-component mixtures in nozzles, orifices and short tubes. J. Heat Transfer, May issue, 179–187 (1971)
5. Y. Munson, Okiishi, Fundamentals of fluid mechanics (underlined) (John Wiley and Sons, Hoboken, NY, 4th Edition 2002)

Chapter 8
Two-Phase Critical Flow

Abstract In the previous chapter, the mass efflux associated with an ideal, perfect gas was derived from six conditions/assumptions and equations for the first and second laws of thermodynamics; this model was then extended to a van der Waal gas. In this chapter, a homogenous equilibrium flow is derived and the effects of a non-homogenous flow is demonstrated. The chapter goes on to derive equations for critical flow associated with non-equilibrium flow where there may be non-equilibrium effects due to momentum, heat transfer, or mass transfer between phases.

Keywords Work of Fauske and Henry · Homogenous equilibrium flow · Frozen flow · Slip velocity · Non-equilibrium critical flow

8.1 Preview

Chapter 7 provide a discussion of critical flow for a purely gaseous phase. Included in Chap. 7 was an introduction to some of the real gas behavior associated with gases obeying a cubic relationship between pressure and specific volume and the chapter went on to introduce a simple model addressing two-phase critical flow behavior—the homogenous equilibrium model.

In this chapter, a more complicated picture is painted and involves two-phase (gas/liquid) [1–6] flow where now there may be non-equilibrium effects between phases in terms of significant heat, mass, and momentum transfer; this chapter will only consider gas/liquid flows with one component. One possible situation is how does slip (liquid velocity does not equal gas velocity) affect the critical flow (flux)? Other possible effects include non-isentropic flow, mass transfer between phases, momentum transfer between phases, and the effects of nucleation on the critical flow.

H. Foust, *Speed of Sound and Critical Flow*, Shock Wave and High Pressure Phenomena, https://doi.org/10.1007/978-981-97-5918-7_8

The chapter begins by addressing two simple modeling paradigms for liquid/gas flows, which are

1. Homogenous Equilibrium Flow
2. Separated Flow.

In homogenous equilibrium flow, the properties between the gas phase and liquid phase are essentially averaged and the first gas/liquid critical flow model was based on this idea, which is called the *homogenous equilibrium model*. A related model that will be presented is the *frozen model*, where the quality between the stagnation point and choke point is assumed to remain fixed.

Another modeling paradigm is to separate the liquid and gas flows where the liquid is a sheath around a gas core and known as annular flow (see Sect. 8.3) also known as separated flow. It has been shown experimentally the separated flow model is accurate for a single component liquid/gas as long as the quality remains above 10%.

The chapter goes on to address non-equilibrium effects [7–14] in order to develop a critical flow model appropriate for low ($x < 10\%$) and very low ($x < 1\%$) quality flows.

In Chap. 9, we'll present a method that has been utilized for homogenous flow for many geometries and extended too many mixtures of liquids and gases. This method is known as the Omega method [15–17].

It should be noted that this chapter assumes a fair amount of knowledge about gas dynamics, multiphase flow, and thermodynamics; for those wishing to further their understanding in this area the following resources are available [1, 3, 13, 18–21].

8.2 Homogenous Flows

The simplest way to model critical flow for a two-phase system is as follows where the following assumptions have been made [11, 13]:

1. No slip (velocity of liquid equals velocity of gas)
2. Thermal equilibrium
3. Isentropic flow
4. Thermodynamic properties associated with equilibrium thermodynamics.

Given the following energy balance

$$h_i + \frac{u_i^2}{2} = h_e + \frac{u_e^2}{2}, \tag{8.1}$$

where $\{h_i, u_i\}$ are the enthalpy and average flow velocity associated with the stagnation point and $\{h_e, u_e\}$ are the enthalpy and average flow velocity associated with the discharge point (choke point) now there are two phases and along with stagnation condition ($u_i \approx 0$) results in

$$x_i h_{i,g} + (1 - x_i) h_{i,l} = x_e h_{e,g} + (1 - x_e) h_{e,l} + \frac{u_e^2}{2} \tag{8.2}$$

And the conservation of mass for a two-phase system at the nozzle is

$$G^* = \frac{u_e}{x_e v_{e,g} + (1 - x_e) v_l}, \tag{8.3}$$

where u_e is the velocity as the choke point, x_e is the quality at the choke point, and $\{v_{e,g}, v_{e,l}\}$ are the specific volume for the gas and liquid components at the choke point.

And for isentropic conditions the following equation applies

$$x_i s_{i,g} + (1 - x_i) s_{i,l} = x_e s_{e,g} + (1 - x_e) s_{e,l} \tag{8.4}$$

From Eq. 8.4, we can determine x_e; from Eq. 8.2, we can determine u_e; and from Eq. 8.3 along with knowing the specific volume, we can determine the critical flux [kg/m^2 s].

Example 8.1 Homogeneous Equilibrium Flow

Given steam at 200 kPa and $x = 90\%$ that converges to 50 kPa, what's the critical mass flux assuming a homogenous critical flow? (Fig. 8.1).

For the *frozen composition flow*, the following assumptions are made:

1. No Slip
2. No heat, nor mass transfer between phases and quality is constant
3. The vapor expands according to $Pv^n =$ constant where n is taken as 1.3
4. All kinetic energy comes from the gas phase
5. Isentropic and ideal gas flow.

Substance = Water

Inlet Conditions

P(i)	[kPa]	200
x(i)		85%
s(I,g)	[kJ/Kg-K]	7.1271
s(I,l)	[kJ/Kg-K]	1.53
s(I,mix)	[kJ/Kg-K]	6.287535
h(I,g)	[kJ/kg]	2706.63
h(I,l)	[kJ/kg]	504.68
h(I, mix)	[kJ/kg]	2376.338

Final Conditions

P(f)	[kPa]	50
x(f)		79.9%
s(f,g)	[kJ/Kg-K]	7.5939
s(f,l)	[kJ/Kg-K]	1.091
s(f,mix)	[kJ/Kg-K]	6.287535
s(f,mix)-s(I,mix)	[kJ/Kg-K]	0
h(f,g)	[kJ/kg]	2645.87
h(f,l)	[kJ/kg]	340.47
h(f,mix)	[kJ/kg]	2132.739
u[e]^2/2	[m^2/s^2]	193598.4
u[e]	[m/s]	622.25

Critical Flow

P(f)	[kPa]	50
x(f)		79.9%
nu(f,g)	[m^3/kg]	3.24035
nu(f,l)	[m^3/kg]	0.00103
nu(f,mix)	[m^3/kg]	2.589604
Critical Flow	[kg/m^2-s]	240.3

Fig. 8.1 Solution for Example 8.1

First Law for an adiabatic, expanding gas is

$$w_{\mathrm{gas}} = h_1 - h_2 \tag{8.5}$$

And Second Law for the system is

$$\Delta s = \frac{\delta q}{T} + \sigma_{\mathrm{sys}} = 0, \quad s_1 = s_2 \tag{8.6}$$

And because the system is isentropic

$$\frac{T_2}{T_1} = \left(\frac{P_2}{P_1}\right)^{\gamma-1/\gamma} \tag{8.7}$$

And substituting Eq. 8.7 into Eq. 8.5 for T_2 results in

$$w_{\mathrm{gas}} = c_{\mathrm{P}} T_1 \left\{ 1 - \left(\frac{P_2}{P_1}\right)^{\gamma-1/\gamma} \right\} = c_{\mathrm{P}} T_1 \left\{ 1 - \left(\frac{P_2}{P_1}\right)^{R_{\mathrm{g}}/c_{\mathrm{p}}} \right\} \tag{8.8}$$

And

$$c_{\mathrm{p}} T_1 \left\{ 1 - \left(\frac{P_2}{P_1}\right)^{R_{\mathrm{g}}/c_{\mathrm{p}}} \right\} = h_1 - h_2 \tag{8.9}$$

Substituting into Eq. 8.9 the following

$$\frac{R_{\mathrm{g}}}{C_{\mathrm{p}}} = \frac{\gamma - 1}{\gamma} \tag{8.10}$$

And

$$C_{\mathrm{p}} T_1 = \frac{\gamma}{\gamma - 1} x_1 P_1 v_1 \tag{8.11}$$

Results in

$$\frac{\gamma}{\gamma - 1} x_1 P_1 v_1 \left\{ 1 - \left(\frac{P_2}{P_1}\right)^{\gamma-1/\gamma} \right\} = h_1 - h_2 \tag{8.12}$$

Conservation of Energy for this system is

$$h_i = h_e + \frac{V_e^2}{2} \tag{8.13}$$

And Conservation of Mass for this system is

$$g = \frac{\dot{m}}{A} = \frac{V_e}{\nu_e}, \tag{8.14}$$

where specific volume (ν_e) is defined as

$$\nu_e = (1 - x_0)\nu_{l,i} + x_0 \nu_{g,i} \eta^{1/\gamma}, \tag{8.15}$$

where

$$\eta = \left(\frac{2}{\gamma + 1}\right)^{\gamma/\gamma - 1} \tag{8.16}$$

Example 8.2 Frozen Equilibrium Flow

Given steam at 200 kPa and $x = 90\%$ that converges to 50 kPa, what's the critical mass flux assuming a frozen composition flow? (Fig. 8.2).

Giver in Fig. 8.3 is the variation of critical flow using HEM and Frozen equilibrium for quality running from 55 to 90%.

Substance = Water

Inlet Conditions

P(i)	[kPa]	200
x(i)		90%
Gamma		1.3
nu(i,g)	[m^3/kg]	3.24034
nu(i,l)	[m^3/kg]	0.00103
nu(i,mix)	[m^3/kg]	2.916409

Final Conditions

P(f)	[kPa]	50
x(f)		90.0%

Critical Flow

P(f)	[kPa]	50
x(f)		90%
h(1)-h(2)	[kJ/kg]	622.8
u[e]^2/2	[m^2/s^2]	622814
u[e]	[m/s]	1116.077
nu(f,g)	[m^3/kg]	3.24035
nu(f,l)	[m^3/kg]	0.00103
nu(f,mix)	[m^3/kg]	2.916418
Critical Flow	[kg/m^2-s]	382.7

$$\frac{\gamma}{\gamma - 1} x_1 P_1 v_1 \left\{ 1 - \left(\frac{P_2}{P_1}\right)^{\gamma-1/\gamma} \right\} = h_1 - h_2$$

Fig. 8.2 Solution for Example 8.2

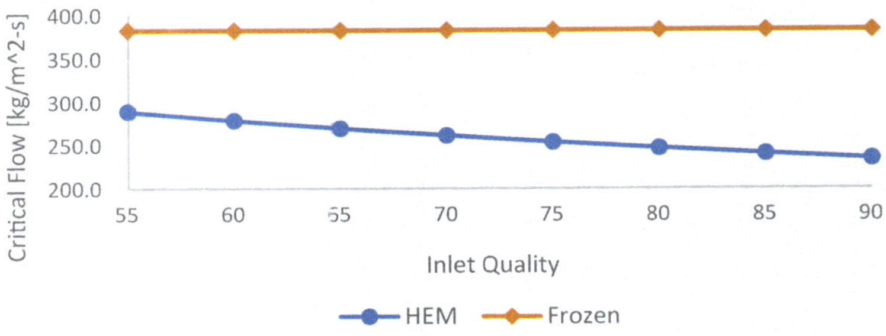

Fig. 8.3 Critical flow versus inlet quality

8.3 Separated Flows

One of the early efforts to address velocity slip was Starkman et al. [11]. In Starkman et al. [11], the energy balance for non-homogenous flow is

$$h_i = h_e + x\frac{u_{e,g}^2}{2} + (1-x)\frac{u_{e,1}^2}{2} \tag{8.17}$$

Equation 8.17 can be arranged to solve for u_g and is given as

$$u_{e,g}^2 = \frac{2(h_0 - h_1)}{x + (1-x)k^2}, \tag{8.18}$$

where h_i is the enthalpy at the stagnation point and h_e is the enthalpy at the choke point and slip (k) is

$$k = \frac{u_l}{u_g} \tag{8.19}$$

The mass balance for this system is

$$\dot{m}_{\text{slip}} = \frac{A_g u_g}{v_g} + \frac{A_1 u_1}{v_1} \tag{8.20}$$

And diving both slides by A

$$\frac{\dot{m}_{\text{slip}}}{A} = \frac{A_g u_g}{v_g A} + \frac{A_1 u_1}{v_1 A} = \frac{u_g}{x v_g} + \frac{u_1}{(1-x)v_1} \tag{8.21}$$

For homogenous flow, we saw the energy balance resulted in

$$u^2 = 2(h_0 - h_1) \tag{8.22}$$

And the mass balance is

$$\dot{m}_{\text{homo}} = A\frac{u}{xv_{\text{g}} + (1-x)v_{\text{l}}} \tag{8.23}$$

And diving both sides by A

$$\frac{\dot{m}_{\text{homo}}}{A} = \frac{u}{xv_{\text{g}} + (1-x)v_{\text{l}}} \tag{8.24}$$

Dividing Eq. 8.21 by Eq. 8.24 results in

$$\frac{\dot{m}_{\text{slip}}}{\dot{m}_{\text{homo}}} = \frac{\frac{u_{\text{g}}}{xv_{\text{g}}} + \frac{u_l}{(1-x)v_{\text{l}}}}{\left[\frac{u}{xv_{\text{g}}+(1-x)v_{\text{l}}}\right]} = \frac{\left[xv_{\text{g}} + (1-x)v_{\text{l}}\right]}{u}\left[\frac{u_{\text{g}}}{xv_{\text{g}}} + \frac{u_{\text{l}}}{(1-x)v_{\text{l}}}\right] \tag{8.25}$$

or

$$\frac{\dot{m}_{\text{slip}}}{\dot{m}_{\text{homo}}} = \frac{\frac{u_{\text{g}}}{xv_{\text{g}}} + \frac{u_l}{(1-x)v_{\text{l}}}}{\left[\frac{u}{xv_{\text{g}}+(1-x)v_{\text{l}}}\right]} = \left[xv_{\text{g}} + (1-x)v_{\text{l}}\right]\left[\frac{1}{xv_{\text{g}}} + \frac{k}{(1-x)v_{\text{l}}}\right]\frac{u_{\text{g}}}{u} \tag{8.26}$$

And

$$\frac{\dot{m}_{\text{slip}}}{\dot{m}_{\text{homo}}} = \frac{\left[xv_{\text{g}} + (1-x)v_{\text{l}}\right]}{\left[xv_{\text{g}} + \frac{1-x}{k}v_{\text{l}}\right]}\frac{u_{\text{g}}}{u} \tag{8.27}$$

In Starkman et al. [11] it is shown that

$$\frac{u_{\text{g}}}{u} = \frac{1}{\sqrt{x + (1-x)k^2}} \tag{8.28}$$

Substituting in Eq. 8.28 into Eq. 8.27 results in

$$\frac{\dot{m}_{\text{slip}}}{\dot{m}_{\text{homo}}} = \frac{\left[xv_{\text{g}} + (1-x)v_{\text{l}}\right]}{\left[xv_{\text{g}} + \frac{1-x}{k}v_{\text{l}}\right]}\frac{1}{\sqrt{x + (1-x)k^2}} \tag{8.29}$$

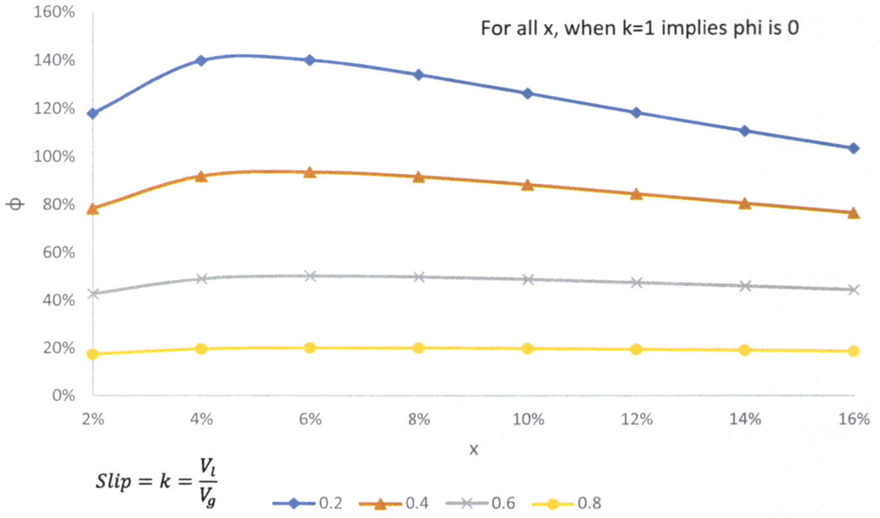

Fig. 8.4 Deviation from homogenous equilibrium flow

Given in Fig. 8.4 is a comparison for the mass flow rate between the developed slip model and homogenous model utilizing Eq. 8.29 where the deviation from homogenous flow is shown as

$$\phi = \frac{\dot{m}_{\text{slip}} - \dot{m}_{\text{homo}}}{\dot{m}_{\text{homo}}} \tag{8.30}$$

It is shown that for even moderate slip ($k = 0.2$) that a homogenous model is not appropriate, and this disparity grows worse at lower quality (see near $x = 4\%$).

Another critical flow model that addresses slip between the gas and liquid phase is as follows [22].

From the conservation of energy and assuming no heat transfer, no work, no appreciable kinetic energy at the stagnation point, nor changes in potential energy then

$$h_i = \frac{u_e^2}{2} + h_e, \tag{8.31}$$

where now u_e is not necessary the local speed of sound, but a mixture of two phases and another form of Eq. 8.31 is

$$2\Delta h = xu_g^2 + (1 - x)u_l^2 \tag{8.32}$$

And it can be show that G, which is the **mass flux**, is equal to

$$G = \rho_l u_l \frac{1-\alpha}{1-x} = Q_l \frac{1-\alpha}{1-x} \tag{8.33}$$

or

$$G = \rho_g u_g \frac{\alpha}{x} = Q_g \frac{\alpha}{x} \tag{8.34}$$

From Eq. 8.34,

$$\alpha = \frac{xG}{Q_g} \tag{8.35}$$

And Substituting Eq. 8.35 into 8.33 results in

$$G = Q_l \frac{1 - \frac{xG}{Q_g}}{1-x} \tag{8.36}$$

And

$$G(1-x) + \frac{xQ_l}{Q_g} G = G\left[(1-x) + \frac{xQ_l}{Q_g}\right] = Q_l \tag{8.37}$$

And solving for G

$$G = \frac{Q_l}{(1-x) + x\frac{Q_l}{Q_g}} = \frac{1}{\frac{x}{Q_g} + \frac{1-x}{Q_l}} = \frac{1}{\frac{x}{\rho_g u_g} + \frac{1-x}{\rho_l u_l}} \tag{8.38}$$

And defining slip velocity (k) as

$$k = \frac{u_g}{u_l} \tag{8.39}$$

It is the case that

$$\frac{2\Delta h}{G^2} = \left[\frac{x}{\rho_g u_g} + \frac{1-x}{\rho_l u_l}\right]^2 \left[x u_g^2 + (1-x) u_l^2\right] \tag{8.40}$$

And substituting in for u_g the term $k u_l$ results in

$$\frac{2\Delta h}{G^2} = \left[\frac{x}{\rho_g k u_l} + \frac{1-x}{\rho_l u_l}\right]^2 \left[x k^2 u_l^2 + (1-x) u_l^2\right] \tag{8.41}$$

And

$$\frac{2\Delta h}{G^2} = \left[\frac{x}{\rho_g k} + \frac{1-x}{\rho_l}\right]^2 \left[x k^2 + (1-x)\right] \tag{8.42}$$

For a given x and dh, G can be maximized by taking the derivative of both sides with respect to k.

$$\frac{\mathrm{d}}{\mathrm{d}k}\left\{\frac{2\Delta h}{G^2}\right\} = \frac{\mathrm{d}}{\mathrm{d}k}\left\{\left[\frac{x}{\rho_{gk}} + \frac{1-x}{\rho_l}\right]^2 \left[x k^2 + (1-x)\right]\right\} = \frac{\mathrm{d}}{\mathrm{d}k} AB = A'B + AB', \tag{8.43}$$

where

$$0 = \frac{\mathrm{d}}{\mathrm{d}k}\left\{\frac{2\Delta h}{G^2}\right\} = 2kx\left[\frac{x}{k\rho_g} + \frac{1-x}{\rho_l}\right]^2 - 2\left[x k^2 + (1-x)\right]\frac{x}{k^2\rho_g}\left[\frac{x}{k\rho_g} + \frac{1-x}{\rho_l}\right] \tag{8.44}$$

And

$$k\left[\frac{x}{k\rho_g} + \frac{1-x}{\rho_l}\right]^2 - \left[x k^2 + (1-x)\right]\frac{1}{k^2\rho_g} = 0 \tag{8.45}$$

or

$$\frac{x}{\rho_g} + k\frac{1-x}{\rho_l} - \frac{x}{\rho_g} - \frac{1-x}{k^2\rho_g} = 0 \tag{8.46}$$

Further

$$k\frac{1-x}{\rho_l} = \frac{1-x}{k^2\rho_g} \tag{8.47}$$

And

$$k^3 = \frac{\rho_l}{\rho_g} = \frac{\nu_g}{\nu_l} \tag{8.48}$$